MW01400964

*Fish and Shellfish
Quality Assessment*

Fish and Shellfish Quality Assessment

A Guide for Retailers and Restauranteurs

Ian Doré

An Osprey Book
Published by Van Nostrand Reinhold
New York

An Osprey Book
(Osprey is an imprint of Van Nostrand Reinhold)
Copyright © 1991 by Ian Doré

Library of Congress Catalog Card Number 91-10901
ISBN 0-442-00206-8

All rights reserved. No part of this work covered by the copyright hereon may be reproduced or used in any form or by any means—graphic, electronic, or mechanical, including photocopying, recording, taping, or information storage and retrieval systems—without written permission of the publisher.

Printed in the United States of America.

Van Nostrand Reinhold
115 Fifth Avenue
New York, New York 10003

Chapman and Hall
2–6 Boundary Row
London, SE1 8HN, England

Thomas Nelson Australia
102 Dodds Street
South Melbourne 3205
Victoria, Australia

Nelson Canada
1120 Birchmount Road
Scarborough, Ontario M1K 5G4, Canada

16 15 14 13 12 11 10 9 8 7 6 5 4 3 2 1

Library of Congress Cataloging-in-Publication Data

Doré, Ian, 1941–
   Fish and shellfish quality assessment : a guide for retailers and restauranteurs / by Ian Doré.
     p. cm.
   Includes bibliographical references and index.
   ISBN 0-442-00206-8
   1. Seafood—Evaluation. 2. Food—Quality. 3. Fishery processing—Quality control.   I. Title
TX556.5.D67   1991
664'.9497—dc20   91-10901
                                                                 CIP

Contents

Preface — *vii*

Acknowledgments — *ix*

List of Illustrations and Tables — *xi*

Illustration Credits — *xvi*

Chapter 1 **When Product Arrives** — 1
  Before the Product Arrives • Product Arrives • Sampling

Chapter 2 **Quality and Freshness— Fresh Fish** — 13
  Fresh Fish • Whole Fish • Dressed Fish • Fillets • Appearance • Steaks • Cook and Eat It

v

vi   Contents

Chapter 3   **Quality and Freshness—Frozen Fish**   29
Fresh and Frozen Seafoods • Packaging • Frost • Clumping • Glaze • Freezer Burn • Rancidity • Gaping • Breaded products

Chapter 4   **Quality and Freshness—Shellfish**   43
Live Molluscs • Shucked Molluscs • Live Lobsters • Crabs

Chapter 5   **Substitutions**   55
Abalone • Clams • Cod, Haddock and Similar Species • Crab • Flounders and Soles • Halibut • Lobsters • Mahi-mahi • Orange Roughy • Perch • Pompano • Salmon • Scallops • Shrimp • Red Snapper • Snappers and Rockfish • Swordfish

Chapter 6   **Short Weight, Inferior Size and Inferior Quality**   79
Short Weight • Inferior Size • Inferior Quality

Chapter 7   **How to Test Product**   89
Standards • Net Weights of Frozen Seafoods • Counts and Uniformity • Cooking

Appendix   95

Resources   101

Bibliography   103

Index   109

Preface

"How do I know that I am getting what I want and what I ordered?" If you are in the foodservice, retail or seafood business, this book helps to answer that question for you. Seafood buyers handle an enormous variety of items, often from a wide range of suppliers. Knowing what is good, what is bad and what is misdescribed or mislabeled is increasingly difficult. While there is no substitute for experience, this book gives you the background and information you need for sound judgement of seafood products.

The theme of this manual is "how to judge the quality." To make the details clear, the book is copiously illustrated with over 100 photographs and drawings, in color and in black and white. All critical distinctions and assessments are illustrated with photographs, drawings or diagrams.

Fish and Shellfish: A Guide for Retailers and Restauranteurs takes you step by step through the receiving and inspection processes. It gives you receiving procedures and tells you how to conduct simple, fair tests on product you receive. It also tells you clearly and in detail what quality points to look for with a wide range of seafood product, from live shellfish to frozen, coated portions.

The book gives quality details for fresh fish, frozen fish (including coated products), live shellfish (molluscs as well as lobsters and crabs) and fresh and frozen shellfish products.

A chapter on substitutions gives you comparison points between many different seafood products that may be passed off as more expensive products: all of these are well illustrated to make the distinctions clear.

Short weight, whether deliberate or accidental, is analyzed and the book explains how you check and test for short weights in different types of products. Inferior size, inferior quality and other aspects of poor quality (and fraudulent) deliveries are examined in detail.

This book is for everyone who has to make judgements about fish and shellfish products: judgements about what to buy, whether the product received is what you ordered and whether it is good and wholesome. Its straightforward format and abundant illustrations clarify and simplify the increasingly complex task of assessing the quality of the seafood you buy.

Acknowledgments

The Florida Bureau of Seafood Marketing made this book possible. I am deeply indebted to Charlie Thomas for making the facilities of his bureau available; to Hank McAvoy for superbly organizing a very complex photo shoot, which went like clockwork; and to Randall Prophet, who took the pictures with great and relaxed professionalism and experience. Hank and Randall between them performed miracles. The Bureau is fortunate to have such skilful people and Charlie was most generous to make them available to work with me on this book.

Taking the pictures is only part of the job. Finding the product and situations to photograph is another. Harry H. Bell and Sons in St. Petersburg, Florida allowed us to disrupt their normal operations and provided unmatched facilities for our photo sessions. I thank Bobby and Harry Bell, as well as their cheerful staff, for their help, forbearance and kindness.

John Marley and John Simson of John Marley and Co., Inc in Tampa were similarly generous in allowing us to interrupt their work while we took photographs. David Smith of S. Felicione Seafood Co., Tampa, provided us with great material and the kindest of receptions. Bill Antozzi of the National Marine Fisheries Service cheerfully offered his expertise and assistance.

Sandra Noel, who has illustrated several books of mine in the past, again supplied the handsome and accurate drawings for this one. George Nardi, New England Fisheries Development Association, very kindly produced a number of pictures of species that had not been available to us in Florida. Richard Lord, with his usual and unmatched expertise, confirmed identification of a number of species where my own identification was less than certain.

I am indebted to Hank McAvoy not only for his work in organizing the photo shoot, but also for reviewing the whole text of the book and making many valuable suggestions, which have definitely made it better. Hank is one of the most knowledgeable of seafood experts. He cheerfully shared his expertise with me to the benefit of this book and of my own education. Any errors that remain in this work are all mine.

Ian Doré

January 1991

List of Illustrations

Figure 1.1   The delivery truck should provide adequate protection for your product.
Figure 1.2   Temperature recorders give independent corroboration that transit temperatures have been maintained properly.
Figure 1.3   Damaged cartons do not provide proper protection for the product.
Figure 1.4   A properly strapped carton resists pilferage.
Figure 1.5   Taping may indicate that the contents have been pilfered.
Figure 1.6   Different straps indicate the carton has been refastened—possibly after some product was removed.
Figure 1.7   This carton has clearly been remarked.
Figure 1.8   Product grid, printed on the carton.
Figure 1.9   Gutted fish should be buried in ice, with the bellies down.
Figure 1.10 Cavitation: ice melts away from the containers, allowing warmer air to circulate.
Figure 1.11 Live shellfish should be stored and transported on pallets so that they are not sitting in fresh water from melting ice.
Figure 1.12 A top layer of nice fillets . . .
Figure 1.13 . . . may conceal poorer product underneath.
Figure 2.1   The skin on the aging grouper (on left) is showing signs of fading and wrinkling. Its eye also shows symptoms of aging.
Figure 2.2   Though still colorful, the dead mahimahi loses much of its brightness and iridescence.

xii List of Illustrations

Figure 2.3   American red snapper *(Lutjanus campechanus)* retains its color and appearance well.

Figure 2.4   The gills of the lower fish are turning brown, indicating it is a little older than the other fish, though both are still very fresh.

Figure 2.5   The red kidney has not been removed from the membrane at the top of the belly cavity.

Figure 2.6   A cleanly eviscerated fish.

Figure 2.7   A good, clean fillet.

Figure 2.8   Another good fillet.

Figure 2.9   These fillets have been badly hacked.

Figure 2.10 A badly gaping fillet.

Figure 2.11 A parasitized fillet—the white worm is marked.

Figure 2.12 Some parasites can grow to unsightly lengths.

Figure 2.13 Parasites can be seen if the fillets are "candled" on a light table, which shows light through the fish.

Figure 2.14 The bone structure of a flatfish.

Figure 2.15 The bone structure of a salmon.

Figure 2.16 The top fillet has been deeply V-cut. Pinbones from the lower fillet have been removed with a straight cut, normally used for snappers.

Figure 2.17 Blood spots and belly membrane mar this fillet.

Figure 2.18 Bruising, blood spots, ragged edges and remnants of skin are among the visible defects of these fillets.

Figure 2.19 The subcutaneous layer on this fillet is silver and shiny, which looks fresh and appealing on display.

Figure 2.20 The subcutaneous layer is oxidizing and turning unattractively brown.

Figure 2.21 Butchering large fish. This one is a tuna.

Figure 3.1   Torn cartons do not protect the product properly.

Figure 3.2   Cartons holding IQF product are especially vulnerable to damage from crushing.

Figure 3.3   This carton is taped, but shows the marks left by the original strapping.

Figure 3.4   Frost inside the bag is an indicator of temperature fluctuations which may damage the product.

List of Illustrations **xiii**

Figure 3.5   Frost or ice on the outside of a carton is an indicator of possible temperature abuse during storage or transportation.
Figure 3.6   Clumping.
Figure 3.7   Lightly glazed, block frozen shrimp.
Figure 3.8   Pink salmon, glazed for proper protection.
Figure 3.9   Pink salmon, glaze removed.
Figure 3.10   Freezer burn.
Figure 3.11   Rancidity.
Figure 3.12   Frost on breaded product.
Figure 3.13   Coating defects.
Figure 3.14   More coating defects.
Figure 3.15   Breaded and "imitation" shrimp, made from same size raw material.
Figure 4.1   Tag for container of live shellfish.
Figure 4.2   Onion bags are frequently used for live shellfish.
Figure 4.3   Bushel boxes provide better protection for fragile soft-shell clams.
Figure 4.4   Oyster liquor should be clear, not opaque.
Figure 4.5   Amounts of oyster liquor vary.
Figure 4.6   Lobsters are measured from the back of the eye socket to the back of the carapace.
Figure 4.7   A berried lobster is a female carrying its eggs externally. Possession of berried lobsters is illegal in the United States.
Figure 4.8   Male blue crab.
Figure 4.9   Female blue crab.
Figure 4.10   Dungeness crabs are measured across the widest part of the top shell.
Figure 4.11   Snow crab. *Bairdii* (top) is larger and more spiny than opilio.
Figure 5.1   Left to right: 31/35 domestic white; 26/30 Chinese white; 21/25 Chinese white.
Figure 5.2   Abalone steak (the missing piece was used for electrophoresis to determine the species).
Figure 5.3   Cuttlefish steaks imitating abalone.
Figure 5.4   Haddock. Note the distinguishing black lateral line.

xiv List of Illustrations

Figure 5.5   Atlantic cod, skinned, showing the silvery subcutaneous layer.
Figure 5.6   Pacific cod, skinned.
Figure 5.7   Atlantic pollock has grayish meat, which is quite white when cooked.
Figure 5.8   Cusk is a large gadoid sometimes used instead of cod.
Figure 5.9   Whiting is smaller and has softer meat than most other gadoids.
Figure 5.10 Assorted cod-like fish.
Figure 5.11 Rock crab and meat.
Figure 5.12 Florida stone crab claw (top) is more clearly marked than the South American claw.
Figure 5.13 Imitation crabmeat.
Figure 5.14 Gray sole (on left) is longer and narrower than winter flounder (on right).
Figure 5.15 Left-eyed flatfish at top, right-eyed flatfish underneath. You can re-assemble fillets to indicate which way the head was. This can help in identifying the fish.
Figure 5.16 Lingcod and halibut steaks.
Figure 5.17 Lingcod and halibut steaks. The lingcod has distinctively spotted skin.
Figure 5.18 Northern lobster *(Homarus americanus)*.
Figure 5.19 Tail fans of spiny lobster (top) and clawed lobster (below).
Figure 5.20 Rock lobster tail (top) compared with slipper lobster (below).
Figure 5.21 Mahi-mahi is also known, unfortunately and confusingly, as dolphin.
Figure 5.22 Yellowtail may be substituted for mahi-mahi.
Figure 5.23 Lake perch.
Figure 5.24 Pompano (top) and permit.
Figure 5.25 Left to right: Sea scallops, bay scallops and calico scallops.
Figure 5.26 A genuine American red snapper *(Lutjanus campechanus)*.
Figure 5.27 Six snappers: Left to right, top to bottom: Silk, gray, vermilion, queen, red, mutton.
Figure 5.28 Top: Spotted rose snapper. Below: American red snapper.
Figure 5.29 Pacific ocean perch (Sebastes alutus). Photo: Donald Kramer, University of Alaska.

List of Illustrations   xv

Figure 5.30   Widow rockfish *(Sebastes entomelas)*. Photo: Donald Kramer, University of Alaska.
Figure 5.31   Yelloweye rockfish. *Sebastes ruberrimus.* Photo: Donald Kramer, University of Alaska.
Figure 5.32   Rockfish fillets.
Figure 5.33   Rockfish fillets, skinned side up.
Figure 5.34   Swordfish. Flesh color depends on feed and does not indicate quality.
Figure 5.35   Swordfish (on right) and shark.
Figure 5.36   Shark on top of swordfish.
Figure 6.1   The 2kg package (top) is actually slightly larger than the 5lb box, because of different glazing methods used.
Figure 6.2   Full box of lobster tails.
Figure 6.3   The same box, minus two tails.
Figure 6.4   Two tails have been removed from this box, without leaving any gaps in the neat arrangement of the tails.
Figure 6.5   One complete inner box of rock shrimp was removed and the top layer of boxes replaced.
Figure 6.6   A heavily glazed lobster tail.
Figure 6.7   Six sizes of shrimp, from 21/25 down.
Figure 6.8   A container apparently full of good quality fillets.
Figure 6.9   The same container showing poorer fillets underneath.
Figure 7.1   Equipment for testing drained weight: an electronic scale and a No. 8 sieve.

List of Tables

Table 2.1   Checklist for quality of dressed fish.
Table 2.2   Checklist for fillet quality.

Appendix

Table A.1   Snapper names in the United States.
Table A.2   Flounder and sole names in the United States.

Illustration Credits

The following pictures are by courtesy of the Florida Bureau of Seafood Marketing. Photographer: Randall G. Prophet:
Figures 1.3, 1.4, 1.5, 1.6, 1.7, 1.8, 1.9, 1.10, 1.12, 1.13, 2.1, 2.2, 2.3, 2.4, 2.6, 2.7, 2.9, 2.11, 2.12, 2.13, 2.16, 2.17, 2.18, 2.19, 2.21, 3.1, 3.2, 3.3, 3.4, 3.5, 3.6, 3.7, 3.10, 3.11, 4.4, 4.5, 5.1, 5.12, 5.23, 5.24, 5.25, 5.26, 5.27, 5.28, 5.34, 5.35, 5.36, 6.1, 6.2, 6.3, 6.4, 6.5, 6.6, 6.7, 6.8, 6.9 and 7.1.

The following pictures are by courtesy of George Nardi, New England Fisheries Development Association: 1.11, 4.2, 4.3, 5.11 and 5.14.

The following drawings are ©, 1991, by Sandra Noel: 2.14, 2.15, 4.6, 5.15, 5.18, 5.19 and 5.20.

Figure 4.1 is by courtesy of Pacific Coast Oyster Growers' Association, Olympia, WA.

Figures 5.30, 5.31 and 5.32 are by Donald E. Kramer, Alaska Marine Advisory Program.

Figure 1.2 is by courtesy of Ryan Instruments, Redmond, WA.

Other photos courtesy of National Marine Fisheries Service and by the author.

Chapter *1*

When Product Arrives

Before the Product Arrives

Provide a copy of your purchase order and product specification sheet to the receiving clerk. The specification should be a simple listing of the main details that appear on the purchase order and the markings that should appear on the product containers. Every shipment should be written up on a receiving report, which should be reviewed before you cut a check for the product. The receiving report should list and report on the main items mentioned in this chapter.

The specification sheet for each product will be the basis of your later and fuller inspection of each shipment. The receiving clerk will ascertain that the product described on the bill of lading and the cartons is the same as the product you ordered.

Product Arrives

Most product arrives on a truck. The truck should be clean and in good operating order. It should not look like the truck in Figure 1.1.

If the product is fresh, the truck does not always have to be insulated and refrigerated, although if it has traveled any distance, it should be. Regulations covering shellfish in many states mandate the use of refrigerated trucks for trips of more than one hour. Refrigeration units should always be in good operating order—they are not just for show.

2 When Product Arrives

Figure 1.1. The delivery truck should provide adequate protection for your product.

Frozen product must always be carried on insulated and refrigerated vehicles. Check the internal temperature of the truck, which should have a thermometer. If you have any doubts or if the temperature is higher than -10°F, ask the driver to demonstrate that his reefer is working. Sometimes, to save fuel, drivers will not use the reefer unit until just before they arrive at their destination. If you have any reason to suspect this, ask for temperature recorders to be carried on subsequent shipments. Temperature recorders (see Figure 1.2) work in different ways, but all are capable of telling you if the temperature exceeded a specified level during the journey.

Check the condition of the outer cartons containing your product. Packaging protects product. This is what it is designed to do and it is very important. Frozen product in torn or damaged cartons will suffer freezer burn (dehydration) very quickly. Damaged cartons also encourage (and may be caused by) pilferage.

Individually quick frozen (IQF) product is particularly likely to suffer from packaging damage, because the product is loose and the cartons are not solidly packed. Figure 1.3 shows cartons containing IQF product: the cartons are bent and dented. Some of them are torn. It is very likely that product inside such cartons will have deteriorated. You or your receiving clerk should either inspect the product before accepting it from the trucker, or at least note the

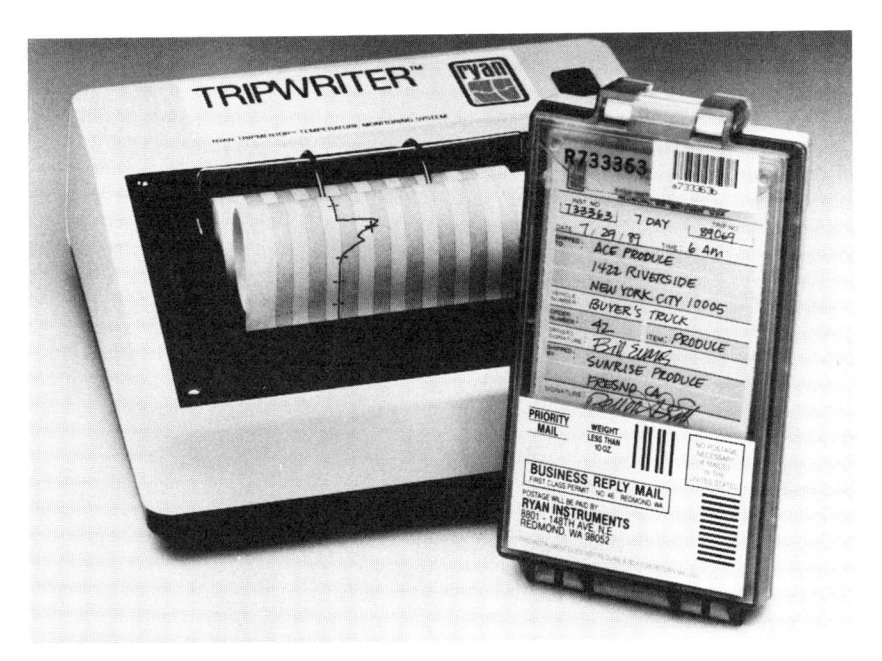

Figure 1.2. Temperature recorders give independent corroboration that transit temperatures have been maintained properly.

damage on the bill of lading and have the driver confirm the problem. It may be the fault of the trucker or the goods may have been handed to him in that condition. Either way, it is not your fault and you should not have to pay for the damage.

Cartons should be solid and well strapped. Figure 1.4 shows a properly strapped carton. Straps cannot easily be replaced, once they are cut. Tape, on the other hand, can easily be undone and replaced. Taped cartons (Figure 1.5) may indicate pilferage. So can restrapped cartons, such as the one in Figure 1.6, where the straps are different colors, indicating that they were put on the carton at different places. Note that U.S. Customs may inspect or take samples of imported product. If they do, the cartons are retaped with special printed tape and the cartons are stamped. If samples have been taken for customs or other inspection purposes, the short weight should be indicated on the carton.

Product descriptions should always be printed or stenciled on the cartons. Obviously, they should agree with the vendor's description and with your purchase order. Look for signs that cartons have been remarked (Figure 1.7) or that the markings have been changed. If you have any doubts, check the contents. Many cartons

4 When Product Arrives

Figure 1.3. Damaged cartons do not provide proper protection for the product.

are preprinted with a grid showing possible contents. The species and size are then checked by the packer. Figure 1.8 shows such a carton. The checks should be made with an indelible marker.

If cartons are packed on a pallet, this should be shrink wrapped. You will sign for each pallet. The trucker is responsible for delivering the sealed pallets, not for the product that is said to be on the pallets. If the pallets are not wrapped, you need to make sure that each one contains the full number of cartons specified. A carton can

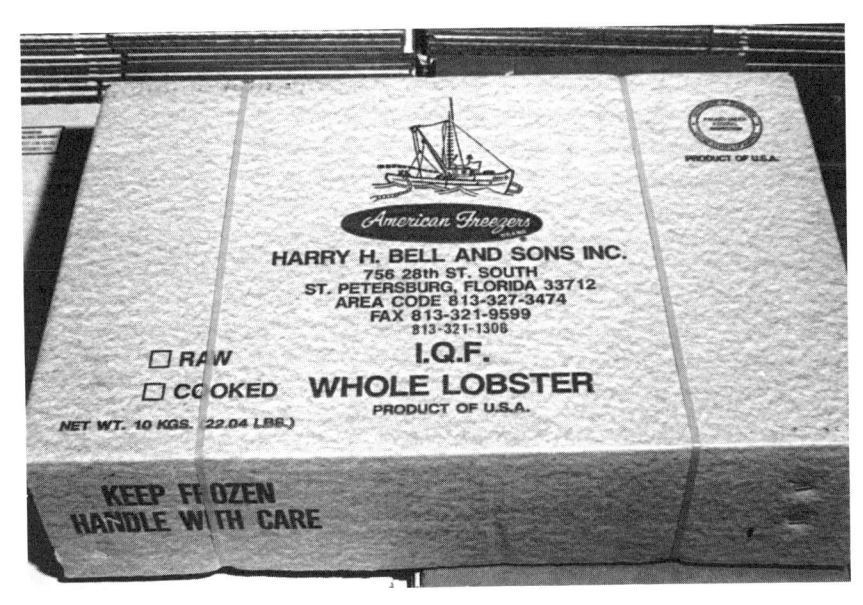

Figure 1.4. A properly strapped carton resists pilferage.

Figure 1.5. Taping may indicate that the contents have been pilfered.

6 *When Product Arrives*

Figure 1.6. Different straps indicate the carton has been refastened—possibly after some product was removed.

Figure 1.7. This carton has clearly been remarked.

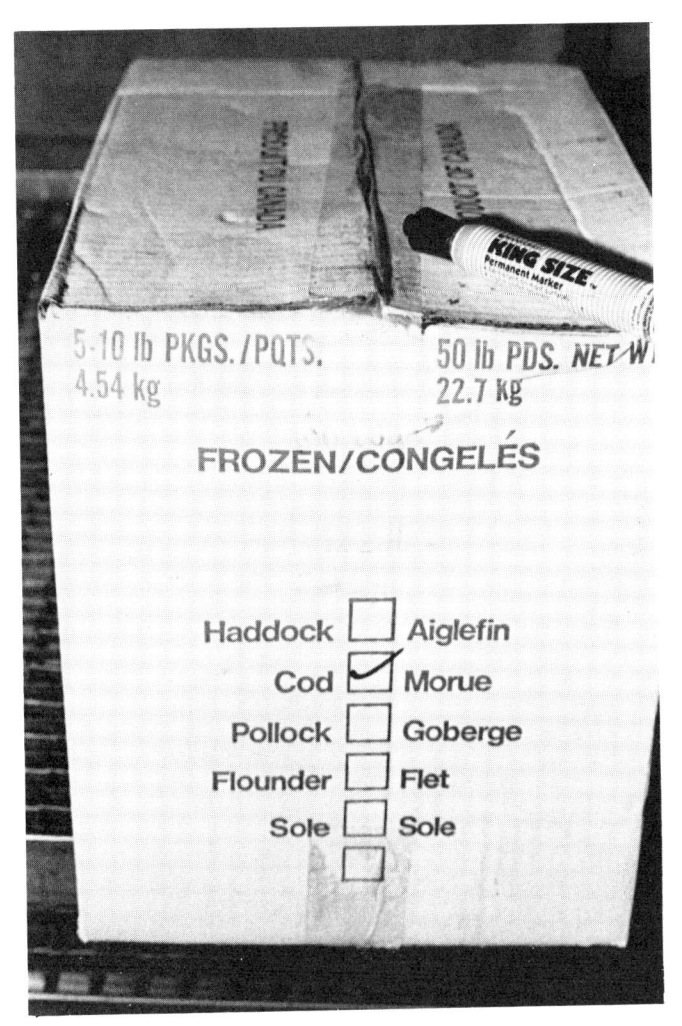

Figure 1.8. Product grid, printed on the carton.

be removed from the center of a pallet without this being apparent from the outside. Theft of a 50 pound carton of shrimp represents a considerable loss. To make sure that it does not happen to you, the safest method is to count every carton on to another pallet. This is laborious and time-consuming. However, once suppliers and truckers know that you do this every so often, they will take better care to ensure that your shipments are complete.

If the shipment consists of several pallets of the same product, you can weigh the pallets. The weights will vary, because of glaze differences and because wooden pallet weights vary, but if one pallet is the weight of a carton less than the others, it makes a lot of sense to unpack and check the light one.

Fresh fish should be properly iced when it is delivered. Figure 1.9 shows whole fish with ice properly packed into the belly cavities and around the fish. The belly cuts should be downwards so that blood and moisture drip away from the fish and do not collect in the cavities. The blood and drip contain spoilage bacteria.

Fresh fillets and shucked shellfish should be in containers with ice packed tightly around them. Figure 1.10 shows how ice melts away from the containers, allowing them to warm up.

Live shellfish, such as clams and oysters, are normally packed in sacks or baskets. Oysters live longer if they are packed with the deep shell down, although this is difficult if they are in sacks. Fresh water can kill live shellfish. If the sacks are iced during transit, the shellfish must not be in contact with the meltwater. Make sure that melting ice has been able to drain away properly from the shellfish. The simplest way to ensure this on a truck is to stack the product

Figure 1.9. Gutted fish should be buried in ice, with the bellies down.

Figure 1.10. Cavitation: ice melts away from the containers, allowing warmer air to circulate.

on pallets so that the water runs away underneath (see Figure 1.11). Clams, mussels and oysters should be packed tightly into their containers: this helps to prevent them from opening and thus dehydrating and dying.

Crabs and lobsters are also unsuited to survival in fresh water and should be checked similarly. Blue crabs should be packed tightly to restrict movement. Both crabs and lobsters should be right side up.

One final point to check before you take product from a truck is that any cooked, fresh product has been carefully segregated from raw or live seafoods. The bacteria on live and raw fish and shellfish tend to multiply very fast if they are transferred to the more appetizing surfaces of cooked product. Make sure that melting ice from fresh product cannot splash or drip on to cooked items. Although these should be in sealed containers, the possibility of cross contamination from dirty ice when the containers are being opened is very real.

Once you are sure that the product is in reasonable condition for delivery, you can accept it from the truck. You should then have a sampling plan and examine the requisite samples from each shipment and type of product.

Figure 1.11. Live shellfish should be stored and transported on pallets so that they are not sitting in fresh water from melting ice.

Sampling

Most of the time, examination of product will be simple, visual inspection to make sure it looks right. If it is fresh product, you will smell it, whether you want to or not. However, opening the first container and looking at the top layer of contents is not a smart way to check your purchases. For one thing, a disreputable supplier might put good product on top of poor product. This technique, illustrated in Figures 1.12 and 1.13, is known as ratpacking. Another reason is that you will not necessarily get a representative sample of a shipment by selecting any container.

Sampling is a statistical technique which you can use to ensure that what you look at very likely represents the entire shipment. Some simple sampling plans are detailed in Title 50 of the Code of Federal Regulations, Paragraph 260.61. These tell you how many cartons or pieces you need to examine to get a reasonably representative sample of the product. When sampling, check whether your shipment is all from a single packing code. If it is not, you will need to sample each lot separately.

Figure 1.12. A top layer of nice fillets ...

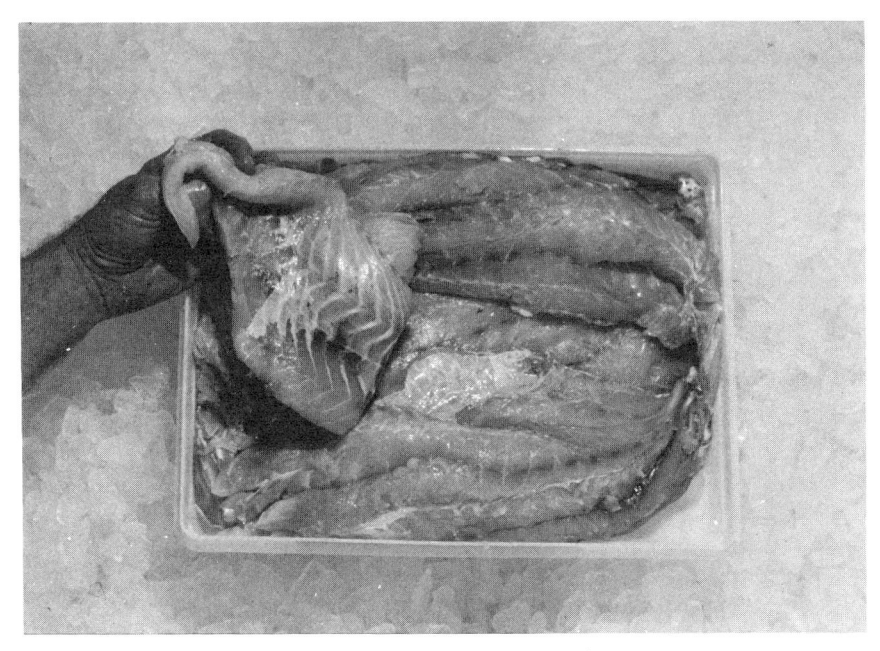

Figure 1.13. ... may conceal poorer product underneath.

Chapter 2

Quality and Freshness—Fresh Fish

Fresh Fish

Fresh fish—product that has not been frozen—demands special care and attention at all stages of handling, buying and using. Some species have very short shelf lives. Others may deteriorate with little visible sign, but still lose flavor and texture. All require constant attention to icing, sanitation and handling.

Frozen product and live product are covered in later chapters. Fresh fish is delivered in a number of forms:

- Head-on, round; that is, with the guts in. This is uncommon except in areas close to fishing grounds, as most fish deteriorates very quickly if left ungutted.
- Dressed; that is, with the gills and viscera removed.
- Fillets, which come in numerous forms, defined and discussed later in this chapter; fillets may be boneless or may contain pinbones.
- Steaks, which are slices of fish cut perpendicular to the backbone. The term also describes chunks of large fish such as swordfish and tuna that do not necessarily meet this definition.

Each of these forms is discussed separately in this chapter. There

Whole Fish

Slime or Mucus

Fish are covered with slime when harvested. In life, the slime helps the fish move through the water and protects it from minor abrasions. Almost all fresh fish are washed free of slime before being shipped to buyers. However, if the slime is present, it should be transparent, almost like water. As the fish ages, the slime becomes milky and then yellowish and thicker. If the slime is almost brown and clotted, the fish is definitely very stale.

Skin and Scales

Skin color varies enormously from species to species and is far from consistent in fish of the same species. However, all fresh fish share the characteristic that the skin looks bright and shiny. The skin should also be tight on the flesh. As the fish ages, the skin becomes dull, loses brightness and color and sometimes bleaches (Figure 2.1). The skin on stale fish may show signs of wrinkling and shrinking away from the flesh.

Skin colors fade after fish is caught. Some fish, such as mahi-mahi, lose their rainbow colors almost as soon as they are taken from the water (Figure 2.2, see color insert). Others, such as some snappers (Figure 2.3, see color insert), retain their color for a long time. You have to be familiar with the characteristics of the species you are buying if you are to make accurate judgments.

The appearance of the skin is greatly affected by the scales. Fish scales vary in size and in how firmly they stay attached to the skin. Salmon have a large number of small scales which give the fish their shiny, silver appearance. These scales are easily rubbed off, leaving the same fish looking dark and much less attractive. Salmon lose scales when they are caught in a net and during handling through packing plants. Farmed salmon may lose scales while still alive, when they are being moved to market. Absence of scales on salmon may sometimes relate to the quality of the fish, but must be interpreted with caution. If the fish has kept almost all

Figure 2.1. The skin on the aging grouper (on left) is showing signs of fading and wrinkling. Its eye also shows symptoms of aging.

its scales, you can be sure that it was handled very gently from the time it was caught.

Fish such as snappers, which have large scales, usually retain them through much washing and abrasion. Again, it is necessary to be familiar with the species you are buying before you can make sound judgements about the individual fish and its condition.

Eyes

The condition of the eyes can be a useful indicator of the freshness of the fish. The eyes of a freshly caught fish will be convex, the pupil black and the cornea (the part surrounding the central pupil) will be translucent. As the fish ages, the eye flattens and eventually becomes concave. The pupil turns gray and sometimes even creamy brown; the cornea becomes opaque and discolored. Figure 2.1 shows how the eyes deteriorate.

Fishermen and processors sometimes pick up fish by the head, using the eye socket to grip it. Such fish will necessarily have sunken and crushed eyes. This handling practice complicates quality evaluation.

Gills

The condition of the gills is sometimes useful in judging freshness. However, in many circumstances, the gills look stale long before any edible part of the fish is even beginning to be affected. Gill color should not be overemphasized as an indicator of fish freshness. The gills of freshly caught fish are bright red, but as the blood in them oxidizes they rapidly turn brownish and any mucus on them turns opaque. Figure 2.4 (see color insert) shows the distinction.

Belly

When fish are not eviscerated, check the bellies for swelling and gas. As whole fish ages after capture, the contents of the gut may ferment and swell. Such deterioration will rapidly spread to the flesh. Small fish that are generally handled whole, such as smelts, should be checked for this condition. To do this, eviscerate a few fish. The edges of the belly cavities should be firm. If the edges of belly flesh pull away with the viscera, this is usually a sign that the guts have started to decompose the fish.

Flesh Along the Backbone

If you split the fish along the backbone and try to lift out the bone, it should stick firmly to the flesh. If the bone lifts out easily, the flesh is stale.

Dressed Fish

Almost all the conditions which apply to whole fish also apply to gutted fish, except, of course, for the condition of the belly. A great deal of gutted fish is also headless, so you do not have eyes and gills as indicators either.

Table 2.1 suggests features to look for and ways to evaluate the freshness and overall quality of dressed fish. It is taken from *Fish Quality Improvement, A Manual for Plant Operators* by Gerald Paquette, published by Van Nostrand Reinhold/Osprey Books. Apart from the gut cavity line, the same factors can be used to rate whole fish.

Table 2.1. Checklist for quality of dressed fish.

	Score 4 (Very good)	Score 3 (Good)	Score 2 (Fair)	Score 1 (Poor)	Enter score
Odor	Fresh, strong, seaweedy, shellfishy	No odor; neutral odor	Definite musty, mousy, bready, malty odor. *Process immediately*	Acetic, fruity, sulphic, faecal. REJECT	
Gut Cavity	Glossy, brilliant, difficult to tear from flesh	Slightly dull, difficult to tear from flesh	Somewhat gritty, somewhat easy to tear from flesh	Gritty, easily torn from flesh	
Gills	Bright red, mucus translucent	Pink mucus, slightly opaque	Gray, bleached, mucus opaque and thick	Brown, bleached, mucus yellowish gray, clotted	
Eyes	Convex, black pupil; translucent cornea	Flat slightly opaque pupil	Slightly concave, gray pupil; opaque cornea	Completely sunken gray pupil; opaque, discolored cornea	
Outer Slime	Transparent or water white	Milky	Yellowish gray, some clotting	Yellowish brown, very clotted and thick	
Skin	Bright, shining, irridescent, no bleaching	Wavy, slight dullness, slight loss of brightness	Dull, some bleaching	Dull, gritty, marked bleaching and shrinkage	

Notes and comments _____

SOURCE: Adapted from Paquette, G., *Fish Quality Improvement, A manual for Plant Operators*. 1983, Van Nostrand Reinhold/Osprey Books.

Belly Cavity

Much more fish is shipped without guts, because of the problems mentioned in the previous section. The gut cavity is an excellent source of information about the condition of the fish. The gut cavity should smell clean and fresh. It should be clean and completely free of traces of blood and viscera. The kidney, which is the dark red organ protected by a membrane along the backbone inside the cavity (Figure 2.5 see color insert), should be completely and cleanly removed (Figure 2.6 see color insert). If this material remains, it can hasten the decomposition of the flesh. The edges of the belly cavity should not be torn. The rib bones should be firmly attached to the flesh of the belly walls. When these bones begin to detach, it is usually a symptom of belly burn. This occurs if the guts are left too long in the fish and is an early sign of decomposition of the flesh around the belly cavity. Belly burn can also begin if the fish is insufficiently cooled and iced.

Fillets

Most of the indicators available to you when examining whole or dressed fish are not available for fillets. Judging freshness and quality involves examination of the condition and workmanship of the fillet. You should also compare the product with the specification, especially if you expect the product to be boneless.

Appearance

Figures 2.7 and 2.8 show good quality fillets. These have shiny, smooth surfaces. Poor quality fillets will show signs of curling at the edges, yellowing of the meat and gaping. As with all seafood products, odor is one of the most important ways you can determine freshness. Ammonia, decomposition and other unpleasant smells indicate that the fish has passed its best.

Fillets should be well trimmed and neatly cut, free of blood spots, pieces of skin or bone and other detritus. A poorly trimmed fillet is shown in Figure 2.9. It is not possible to generalize about flesh color, because each species has a different characteristic. Cod and haddock are white, bluefish is gray (almost blue), while salmon is pink or red. The normal flesh color of each species is

Figure 2.7. A good, clean fillet.

different and the color of individual fish of the same species can vary greatly. Salmon, in particular, undergoes great changes as it matures and swims upstream to spawn. A chum salmon from the ocean may have red flesh, which fades as the fish enters fresh water and can become grayish-white when the fish is close to spawning. (See *Salmon—The Illustrated Handbook for Commercial Users* by Ian Doré, published by Van Nostrand Reinhold, 1990, for full details about salmonid flesh colors as well as actual color guides.)

You become accustomed to the color and general appearance of fillets that you buy regularly; this experience helps you evaluate what you receive.

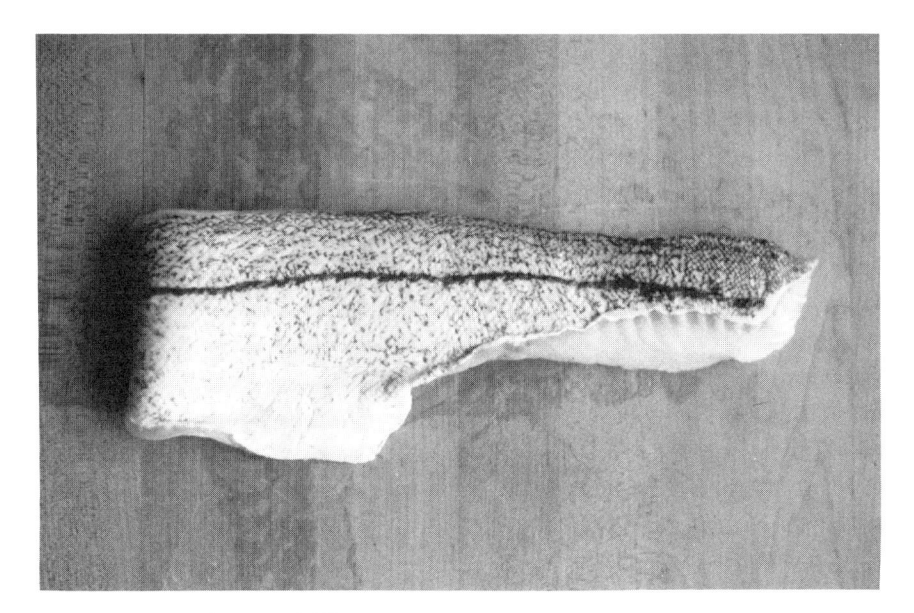

Figure 2.8. Another good fillet.

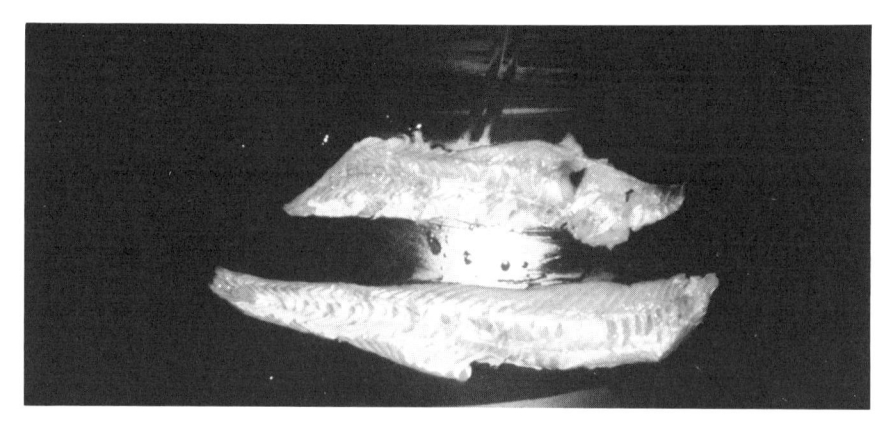

Figure 2.9. These fillets have been badly hacked.

Gaping

Fish flesh is composed of flakes which are held together by a membrane between each flake. The size of the flake is characteristic of each species: cod has a very large flake while flounders have very small flakes. The strength of the connecting membrane determines how well the flakes hold together. Some fish have weak membranes and the flakes separate—gape—very easily. Bluefish and hake are examples of fish which gape readily. Gaping is also dependent on how fish are handled. In some circumstances, incorrect handling can cause a fillet to gape excessively. Most fillets will gape more as they age. Processing fish before it has passed through *rigor mortis* causes fillets to gape excessively, because the fillet passes through rigor without the structural strength of the skeleton to hold it in place: it therefore tears the connecting membranes away from the flakes of the flesh. Fillets will gape more at certain times of year (for example, when the fish have recently spawned and are in poorer condition), so while gaping is an important feature, its evaluation requires some knowledge of the fish and of the way it may have been caught and handled. For a full discussion of gaping and the reasons for it, see *The New Frozen Seafood Handbook* by Ian Doré, published by Van Nostrand Reinhold/Osprey Books.

Once you are familiar with the "normal" amount of gaping you can expect from a particular species of fish fillet, then an abnormal degree of gaping may be cause for concern. Figure 2.10 shows a badly gaping fillet.

Figure 2.10. A badly gaping fillet.

Parasites

Many fish species are hosts to parasites, mostly small worms (Figure 2.11), some of which may be found in the flesh. Almost all fish parasites are harmless to humans and are readily killed in normal cooking. However, all parasites are unsightly: no consumer will relish buying a piece of fish with visible parasites in it. Few would actually risk eating such a product. Parasites are increasingly important not because of their appearance, but because more people are eating their fish raw (for example, in sushi). While it is very rare, certain live parasites in raw fish can infect people. This is why most experts recommend using only commercially frozen fish for sushi: freezing kills most parasites in a couple of days.

Tuna and salmon are often parasitized. Cod frequently has codworms. Black drum has a reputation for being universally and heavily infested. Swordfish may have very large worms in the muscle. Fish from some areas may be heavily infested while the same species from elsewhere may be completely free of parasites. Codworms have increased in numbers markedly in recent years, apparently because seals are now protected and are multiplying. The codworms's life cycle requires the creature to be resident for a while in the intestines of a seal. The increase in the numbers of seals increases the chances of cod fillets having codworms.

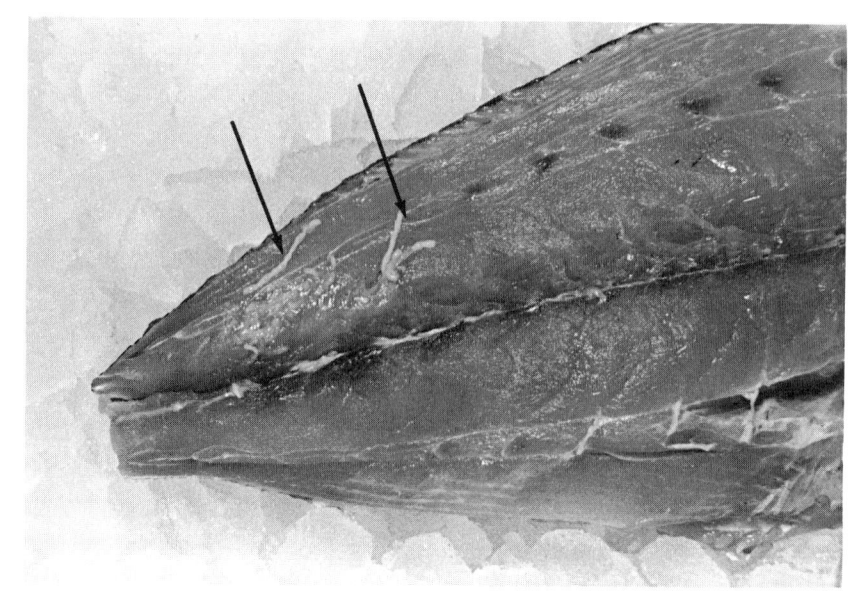

Figure 2.11. A parasitized fillet—the white worm is marked.

Figure 2.12 shows a parasite pulled from the fish. Parasites can be removed easily with a knife. Processors of cod and similar fish, where parasites are expected, pass the fillets over a light table (Figure 2.13), which reveals any worms. These are then removed before the product is packed and shipped. Because most processors understand the need for and the benefits of this process (called candling), the chances of receiving parasitized fish are greatly reduced. However, if you are selling fish retail or serving it raw, parasites are an important concern and you should double check for their presence in fillets.

Bones

Bones are a cause of many misunderstandings and disputes between fish processors and their customers. A fillet is generally defined as the muscle taken from the side of a fish. this does not mean that it is boneless. Many fish species contain a row of small pinbones that are found along the lateral line. Figure 2.14 shows a flatfish, which does not have pinbones. Figure 2.15 shows a salmon, which has a bone structure similar to that of most round-bodied species and has numerous pinbones (numbers vary from species to species). In some species, these pinbones are not attached to the rest

Appearance 23

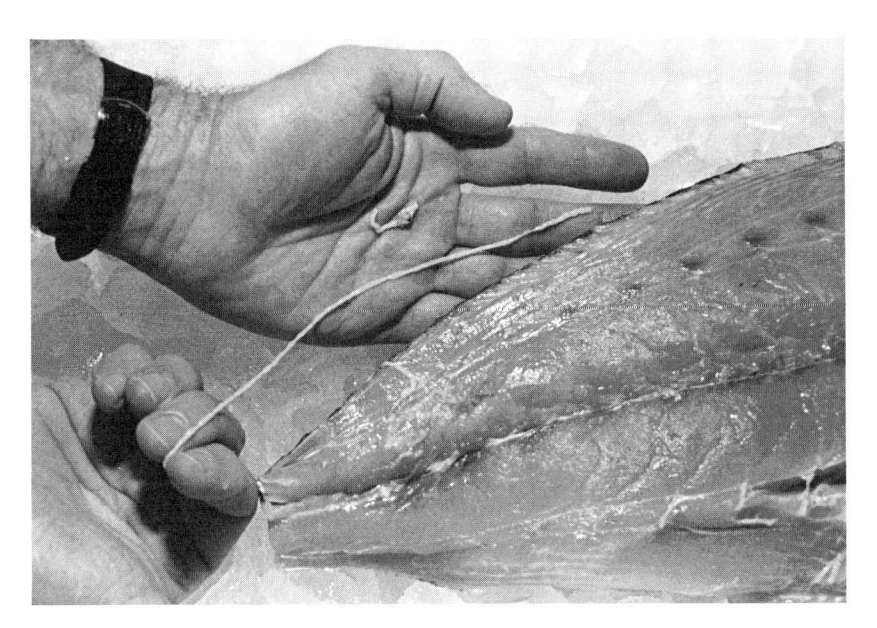

Figure 2.12. Some parasites can grow to unsightly lengths.

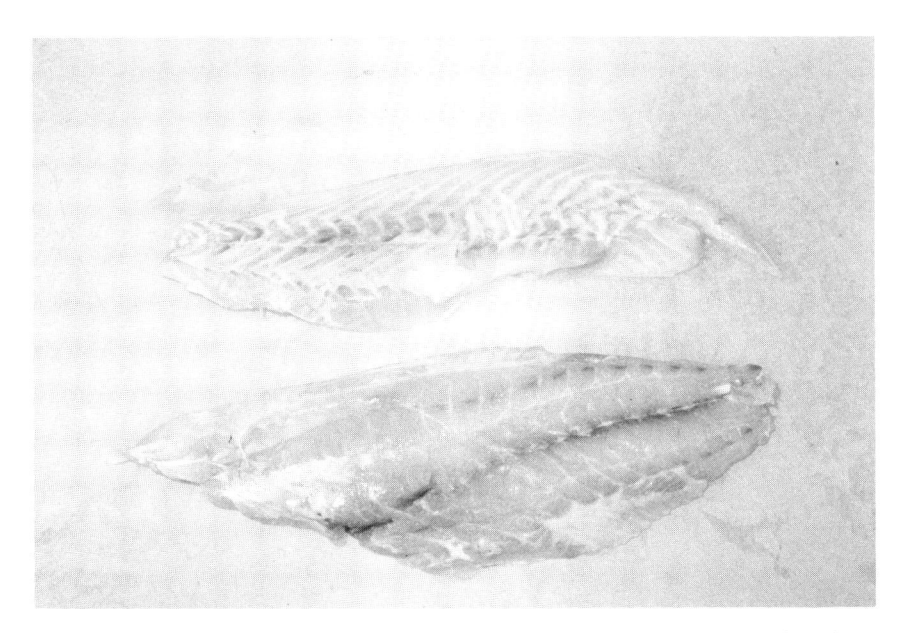

Figure 2.13. Parasites can be seen if the fillets are "candled" on a light table, which shows light through the fish.

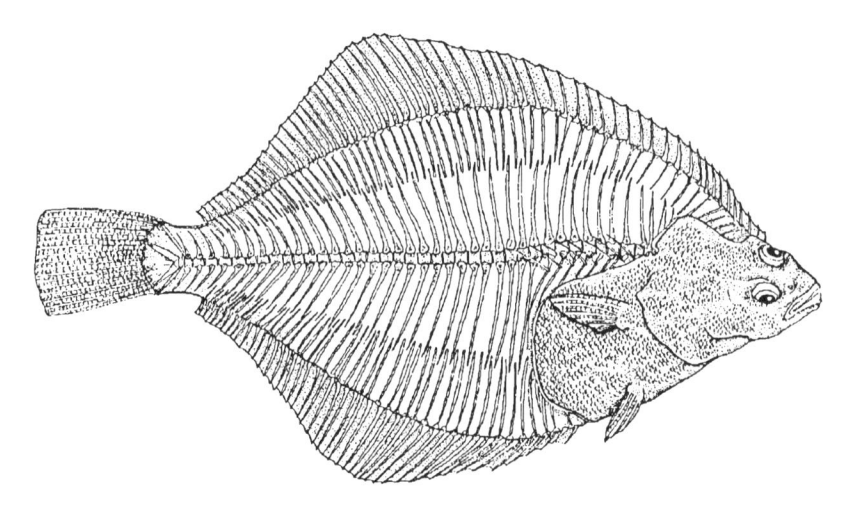

Figure 2.14. The bone structure of a flatfish.

of the skeletal structure, but are free floating in the muscle. This makes them particularly hard to remove. If you have bought boneless fillets, you can expect the fillets to have been fully boned. However, bear in mind that U.S. Department of Commerce standards for bonelessness do permit a few small occurrences: it is virtually impossible to process fish so that every single piece of bone is definitely removed.

To check for unwanted pinbones, feel along the thickest part of the fillet from the head end towards the back. Apply slight pressure and you should be able to feel if there are any bones left in the fillet. There is an excellent and amusing article on bone structures of fish in *Seafood Leader* magazine of March/April 1990.

Different fish are cut in different ways to remove the pinbones, depending on the bone structure of the fish and its cost: it is easy to

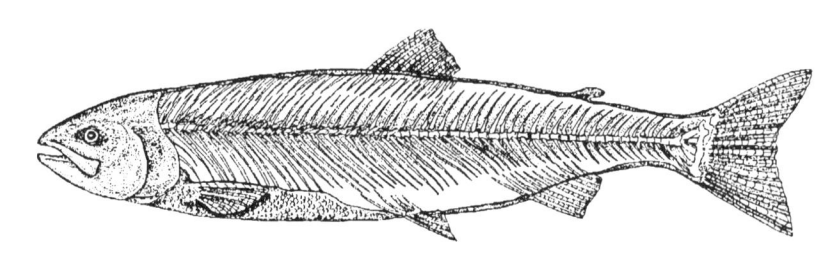

Figure 2.15. The bone structure of a salmon.

remove the bones with a lot of flesh, but this reduces the yield and so increases the price of the fillet. Figure 2.16 shows some of the alternatives. Some fish are cut in a straight line to avoid a fairly short line of pinbones. If these bones extend further down the fish, a J-cut may be used to remove them together with some of the thinner parts of the belly wall. The most labor-intensive cut is the V-cut, where the bones are removed with a strip of flesh. In expensive fish, such as salmon, it is sometimes possible to remove the pinbones individually with pliers, but this method is expensive and not always totally effective.

Processing Defects

Fish are cut and trimmed by machine and by hand, but since neither the fish nor the filleter can be totally consistent, quality of workmanship can vary. Look for bone fragments, blood clots (see Figure 2.17) and parasites that have not been properly removed. Also look for ragged edges, torn fillets and other signs of poor processing (Figure 2.18, see color insert).

Skinning

Many fillets are sold skinless. Fish that have soft flesh and gape readily (such as bluefish) are rarely skinned because the fillets tend to fall apart. However, many popular species are sold without skin.

Figure 2.16. The top fillet has been deeply V-cut. Pinbones from the lower fillet have been removed with a straight cut, normally used for snappers.

Figure 2.17. Blood spots and belly membrane mar this fillet.

Cod and flounder, both mainstays of the retail fish trade, are invariably skinned. The subcutaneous fat layer that lies between the skin and the flesh of many species should be silvery and shiny, not dull and brown: the fat oxidizes as the fish get older and this change can be a sign of aging fillets (Figures 2.19 and 2.20, see color insert).

Skin should be completely removed; there should not be any fragments left on a skinless fillet. In some species, the subcutaneous fat should also be removed (catfish and whiting are examples), because the fat is strongly flavored and detracts from the edibility of the fish.

Table 2.2 shows features to check when receiving fillets and suggests ways to evaluate them for freshness and quality.

Steaks

Evaluate steaks in terms of appropriate features discussed for whole and dressed fish and for fillets. Skin and flesh appearance and cleanliness are important. So, normally, is uniformity in thickness and size. If there is too great a variation between steaks, then cooking times will vary, making them much less convenient to use.

Table 2.2 Checklist for fillet quality.

Feature (Score)	Very good (4)	Good (3)	Fair (2)	Poor (1)
Appearance—general				
gaping				
Odor				
Trim—general				
blood spots				
bruises				
membrane				
skin (fragments on skinless fish)				
Parasites				
Excessive bones or bone fragments				
Dehydration				
Cooked sample:				
Odor				
Flavor				
Texture				

Notes and comments: _____

Large fish, such as swordfish and tuna, are often steaked. However, retailers and restaurants normally buy large chunks of such fish, which may be labeled as bullets, loins, wheels, sides, slabs or some other suitable description. Figure 2.21 shows a tuna being cut into wheels. Center cuts and boneless pieces are normally more expensive because there is less waste. Agree upon the description with your supplier and then check the product against that specification.

Cook and Eat It

The best and most reliable way to make sure that fish is good and wholesome is to cook and eat it. To make a fair assessment, cook a small portion of the meat in a boilable plastic bag, without

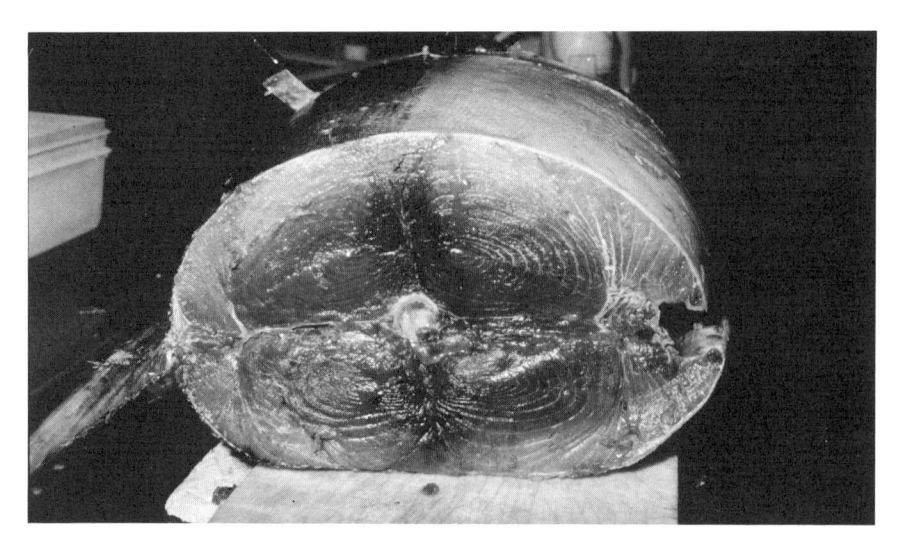

Figure 2.21. Butchering large fish. This one is a tuna.

seasoning. For a more elaborate test, you can bake fish in the oven according to the Canadian cooking method (ten minutes per inch of thickness at 450°F), again without using any seasoning.

Chapter 3

Quality and Freshness—Frozen Fish

Fresh and Frozen Seafoods

The use of the term "fresh" continues to bedevil the seafood business. The word is used as the opposite of stale, as well as to distinguish unfrozen product from frozen. "Fresh" is NOT the opposite of "frozen"; it is simply a description of a different state. Frozen fish must be fresh, meaning that it must not be stale. Read this chapter, indeed all of this book, in that context.

Frozen product is offered in an even greater variety of forms and specifications than unfrozen product. Whole fish, dressed fish, fillets and steaks are all available. So is breaded and battered fish, often cut to exact portion sizes. Frozen prepared meals are common. In this chapter we concentrate on the main commercially used forms and outline things to look for which may indicate quality problems. Factors such as workmanship, gaping and others discussed in the previous chapter on fresh fish all apply to frozen fish also. Read and use those sections when examining frozen fish. Parasites are normally destroyed by freezing and frozen storage of more than a few days, so they are not much of a problem with frozen seafoods.

The rest of this chapter covers some of the aspects of quality which are relevant to frozen product specifically.

Packaging

This is an important and sometimes neglected attribute of frozen fish products. Frozen storage at temperatures around 0°F is intensely dehydrating. Moisture is removed from everything in the store. Seafood, like most other foodstuffs, is destroyed by dehydration and so must be protected from the effects. Two protective methods are used, usually in combination: glazing and packaging.

Packaging is the outer line of defense against the ravages of cold. Good packaging prevents the circulation of air over the surface of the product and by doing this protects the moisture in the surface layers of the product. Plastic shrink wraps, which mold themselves to the shape of the product, can be used for fillets, shellfish meats and other boneless items, but can be punctured by bones or hard frozen surfaces. It is more common to wrap product in plastic and then put it in a box. These inner boxes are placed in an outer carton, or master carton. This gives three layers of protection and, if done properly, is most effective.

Check to ensure that all packaging is tight and unbroken. Torn boxes let air circulate around the product. Torn outer cartons often contain torn or damaged inner boxes (Figure 3.1). Check everything thoroughly to make sure that both outer and inner packaging have maintained their original integrity.

Block frozen product fits tightly into outer cartons and normally stacks easily. Consequently, the packaging is much more likely to be in good shape than it is with IQF product, which is much harder to pack. IQF shrimp, for example, is packed in polyethylene bags, which are placed in the outer cartons. The contents are quite loose, so it is easy to damage cartons when they are stacked: it is like stacking irregularly shaped rocks. Some IQF product is packed in bags and then in inner boxes. This helps to give a little more structural strength to the outer carton and also adds a layer of protection against dehydration. Figure 3.2 shows the effects of poor handling on a stack of boxes of IQF product.

As mentioned in the first chapter, cartons must be properly sealed, both to protect the contents from dehydration and to assure you that all the contents are there. Some packers use as many as four bands around a standard carton, two in each direction. Different shapes of carton may require different strapping formulas: the rules are that there should be plenty of strapping to

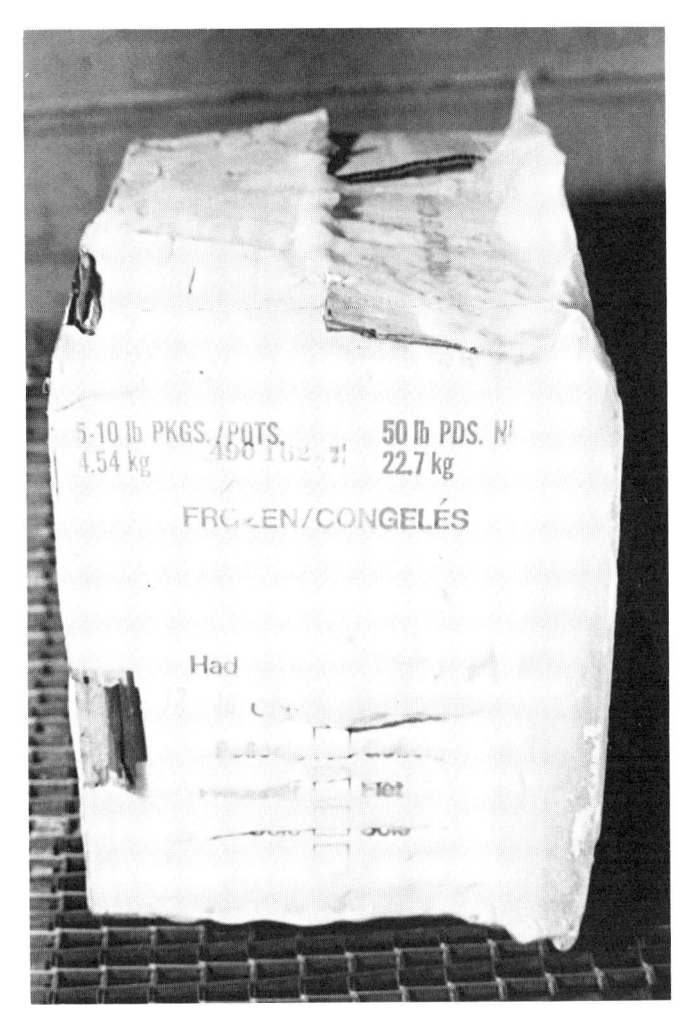

Figure 3.1. Torn cartons do not protect the product properly.

ensure that the carton cannot fall apart and to ensure that anyone trying to pilfer product will have to cut at least one of the straps to get inside the carton. Check that cartons are properly and consistently sealed and strapped. Figure 3.3 shows a taped carton which bears the marks of its original strapping, indicating that it has been opened and resealed. If you have doubts about any carton, open it and check the condition and the weight of the contents.

The most important piece of packaging for product protection is

Figure 3.2. Cartons holding IQF product that are especially vulnerable to damage from crushing.

probably the inner plastic lining or the polybag, whichever is used. This should be good quality plastic and thick enough to give adequate protection. Because there are so many different plastics, many designed to allow the passage of certain gases or to inhibit the passage of others, it is impossible to tell easily whether any particular packaging is adequate. Over time, you can tell whether one product has a better storage life than another and adjust purchasing accordingly.

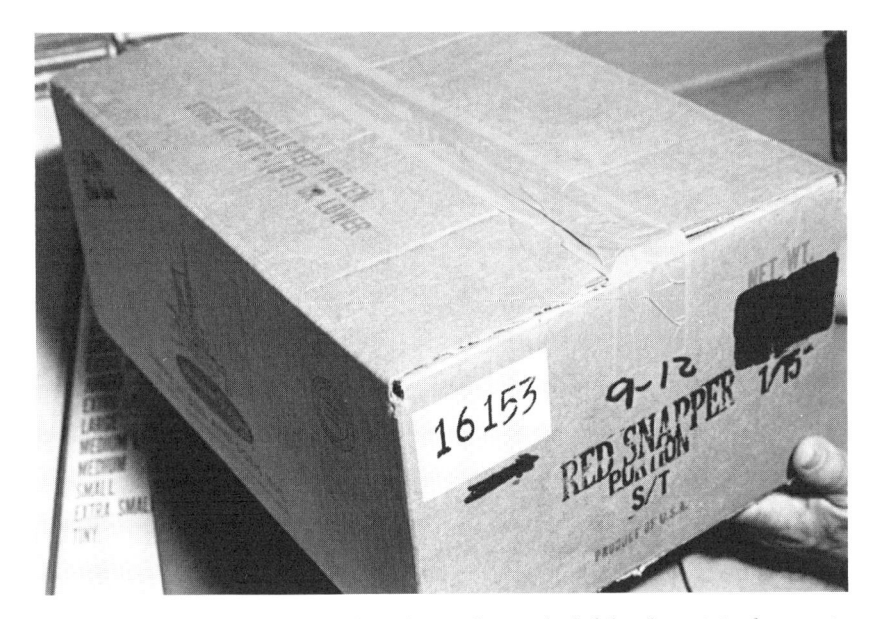

Figure 3.3. This carton is taped, but shows the marks left by the original strapping.

Frost

Frost on or in packages of frozen seafood is invariably a sign of something wrong. Frost—loose particles almost like snow—inside individual bags indicates partial thawing and refreezing (see Figure 3.4). Product which has been subjected to such treatment will have much shorter storage life. It is also quite possible that its texture and flavor will have been affected for the worse. Remember that the moisture that produced the frost all came from the product within the sealed bag.

Frost on the outside of packages is also a sign of something wrong. Figure 3.5 shows an example. Again, it probably indicates temperature fluctuations inside the warehouse or the truck: moisture from warmer air freezes on the colder surfaces of the frozen cartons or product. Warehouse temperatures are supposed to remain quite stable, but some freezers, especially older ones, suffer very wide temperature fluctuations. Frost on and in packages can be a symptom of such a problem.

Ice on packages is another warning signal. If water has been spilled on to a package and frozen, you should inquire where the ice came from and why.

Figure 3.4. Frost inside the bag is an indicator of temperature fluctuations which may damage the product.

Clumping

Figure 3.6 shows clearly what clumping is: frozen product that has been partially (or even totally) thawed and then refrozen. IQF product sticks together in clumps. Some product is very susceptible to clumping: frozen peas, for example, because of their very small size and individual light glaze, will clump at very small changes in temperature. So will very small peeled shrimp. In general, however, clumping is a sign of mishandling. Product showing signs of clumping should be examined very carefully before it is accepted.

Glaze

Almost all frozen product is glazed, with the obvious exception of battered and breaded items. Glaze is a thin coating of ice which covers the exposed surfaces of the seafood so that any dehydration will be of the glaze, not the product. Proper glaze is essential if frozen product is to have a reasonable storage life. Excessive glaze and improper weight claims made on the basis of glazed weights

Figure 3.5. Frost or ice on the outside of a carton is an indicator of possible temperature abuse during storage or transportation.

are a continual problem for all sectors of the seafood industry and its customers. This topic is discussed in more detail in Chapter Six.

Protective glaze is applied according to the needs of each product.

Block Frozen Shrimp

Shrimp is placed in a plastic lined inner box. The shrimp is weighed, the plastic is folded over the product and the box is filled with water and then frozen. This provides a thick protective covering of ice

Figure 3.6. Clumping

which settles firmly around the irregularly shaped product. Some blocks of shrimp have as much as one third of their total weight in glaze. There is absolutely nothing wrong with this, provided that the box contains the full weight of product as well (see Chapter Six). Figure 3.7 shows two boxes of quite lightly glazed shrimp. Similar techniques are used to glaze blocks of fish fillets.

Whole and Dressed Fish

These are usually individually glazed, because their shape makes it difficult to glaze them in any other way. After the fish is frozen, it is dipped in a bath of ice-cold water. A layer of ice immediately freezes on to the cold fish, protecting it from the air. This process is usually repeated two or three times to build up a satisfactory layer of glaze. If salmon is kept in storage for more than three or four months, it will usually be reglazed to restore the protection that has been lost to dehydration in the meantime. Glazed fish is sometimes put into individual polybags for additional protection.

IQF Shrimp

Small product is commonly glazed by passing it through a spray or glazing tank after it has been frozen. Like the whole fish, it picks up

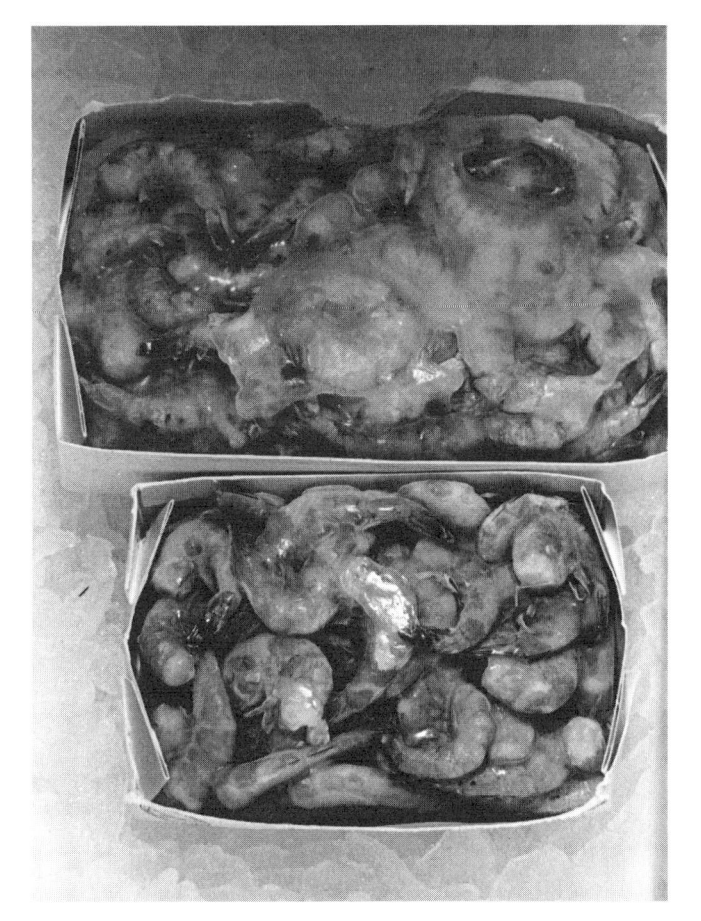

Figure 3.7. Lightly glazed, block frozen shrimp.

a layer of ice; this process can also be repeated as often as necessary to give the required protection. Small product, such as IQF shrimp, have proportionally greater surface area to weight than larger product. Consequently, the percentage of glaze used in a given weight pack of small product will tend to be higher than on larger pieces.

Figures 3.8 and 3.9 show properly glazed product beside unglazed product.

Freezer Burn

Freezer burn is what happens to poorly packaged frozen product. Freezer burn is dehydration and can be recognized easily by whitish

Figure 3.8. Pink salmon, glazed for proper protection.

Figure 3.9. Pink salmon, glaze removed.

or yellowish areas which, in extreme cases, become spongy and soft, even while the product is still frozen hard. Freezer burn affects exposed areas first, then edges and thinner parts. The edges of belly flaps of whole fish are particularly vulnerable. Figure 3.10 (see color insert) shows freezer burned product.

Freezer burn happens to poorly packaged product kept in store for too long. Many items, if well packaged and glazed, can be stored for years, without dehydrating.

Rancidity

This happens naturally and inevitably, but is much hastened if product is poorly packed and stored. Rancidity, which is oxidation of the oils in the flesh, is revealed by yellowish spots on the surface. Figure 3.11 (see color insert) shows a very rancid fish. Rancidity depends partly on the nature of the product: oily fish, such as salmon and tuna, will go rancid long before low-fat fish, such as whiting and cod. Good packaging will help to slow rancidity. One cause of rancidity is fluctuations in temperatures in frozen storage. Although the fish may have remained frozen hard throughout its stored life, sometimes it may suffer wide fluctuations in temperature. This may be because it was stored too close to doors that are frequently opened, or because the freezer was sometimes taking in product that was not sufficiently cold, or simply because the freezer is old and does not maintain stable, low temperatures. One effect of subfreezing temperature fluctuations is to "pump" oil from the tissues to the surface of the fish. Once at the surface, where it is exposed to air, the oil rapidly oxidizes.

Although advanced rancidity is visible on the surface of the fish, lesser oxidation is hard to detect except by cooking and eating a piece of the fish. Cook a small portion in a boilable bag without seasoning, until it is just done. Rancidity can be recognized by a sour taste, uncharacteristic of the fresh fish. Rancid fish can be quite unpleasant to eat. It is worth paying attention to the potential problem when you are buying oily fish.

Gaping

This was covered in the previous chapter. Everything mentioned there applies also to frozen fish. An additional problem with frozen fish occurs if certain species are processed and frozen before passing through *rigor mortis*. They will continue to go through the rigor process even in the frozen state. Frozen fillets may look fine, but may gape and even fall apart when they are thawed or cooked. Whole or dressed fish is less of a problem because the bone structure holds the flesh together.

Breaded products

Look for frost as a sign of temperature abuse. Check that the breading is complete, with no holes, gaps or damaged areas. Check that the product meets specifications for the percentage of coating.

These three areas are the most critical when dealing with breaded and battered seafoods. Of course, it is also wise to check that the fish inside the coating is wholesome and fresh. Do this as you would any other product: cook it, smell it and, if it smells acceptable, eat it.

Frost on breaded product (Figure 3.12, see color insert) is often a sign that the temperature has fluctuated during storage or transportation. It is a particular problem with product that is intended to be fried, because frost turns to steam in hot fat or oil and spatters. Check that inner packages are tight and that all boxes are in good condition so that product is thoroughly protected.

Gaps or holes in breading can also cause oil spatter and even make the breading blow off the surface of the seafood. This not only makes the product unappetizing, it puts burned coating in the oil, which then has to be discarded. Figures 3.13 (see color insert) and 3.14 show some somewhat extreme coating problems.

Good packaging is especially important for coated products, partly because these cannot be glazed to prevent dehydration and partly because coatings, especially the breading types, are often fragile and easily damaged. Adequate packaging to protect coated product is absolutely essential.

The Food and Drug Administration has Standards of Identity in Chapter 21 of the Code of Federal Regulations for breaded shrimp

Figure 3.14. More coating defects.

and for product labeled as "lightly breaded shrimp." These Standards require that at least 50 percent of breaded shrimp consist of shrimp; for lightly breaded, 65 percent of the total weight must be shrimp. The U.S. Department of Commerce's National Marine Fisheries Service has standards for its inspection services. The standards covering coated products also require a minimum of 50 percent seafood inside the coating.

Coated product that does not conform to the 50 percent requirement must, by law, be labeled as "imitation." Figure 3.15 shows the same size shrimp with substantially different amounts of breading. The FDA also requires that products made from surimi which mimic real products be labeled "imitation." Presumably, if anyone bothered to make a surimi-imitation shrimp and then coat it so that the breading percentage was over 50 percent, the product would have to be labeled "imitation imitation shrimp." So far as we are aware, no such product yet exists.

The coating percentage is very important because the shrimp, fish, oysters or scallops inside the coating are much more expensive than the coating. Testing the percentage of coating is quite easily done, following procedures set out in the U.S. Grade standards (Title 50, U.S. Code).

Coated products constitute a huge proportion of the total seafood

Figure 3.15. Breaded and "imitation" shrimp, made from same size raw material.

business in the United States. Most of them are made by very large companies that have careful quality control and excellent reputations, which they take great pains to protect. Problems with coated products will almost all be with unbranded or unknown-label items, which often sell at an apparently good discount. Adding breading is a simple way to make a product that costs less overall, but that does not mean that the product is a better buy.

Chapter 4

Quality and Freshness—Shellfish

Live Molluscs

In terms of quantity as well as value, clams and oysters are the most important molluscan shellfish sold live in the United States. With mussels, their harvest and distribution is controlled under a federal program called the National Shellfish Sanitation Program (NSSP). Producing and consuming states are responsible for administering the program, which has been in operation since 1925.

Clams, oysters and mussels are filter feeders. This means that they pump water through their systems and extract the nutrients they need. Clams and oysters are often eaten raw. All three animals are often eaten lightly cooked. We also eat the entire animal, not just the "flesh," as we do with fish. All of this means that any harmful organisms that are in the water can be transmitted to human consumers through these shellfish. Such harmful organisms include bacteria and viruses. Shellfish can also concentrate toxins produced from certain algae that the shellfish eat. Some of these toxins, popularly known as "red tide" poisons, can be extremely dangerous. Shellfish growing waters are monitored for signs of the causative algae and closed if necessary. The NSSP was originally established because of outbreaks of typhoid from shellfish harvested from water polluted with untreated sewage. It remains in place to ensure that shellfish are harvested only from approved waters and that the distribution of these animals is done under sanitary conditions.

The program has been successful and there are remarkably few

43

incidents of illnesses from shellfish, although those that do occur are often given headline treatment. It would be wrong to pretend that there are no risks: shellfish harvested from polluted water can be very dangerous. Everyone in the industry or concerned with selling or serving shellfish should be aware of the rules and cooperate in following them. This is not the place for a detailed discussion of how the NSSP works; only the basic rules that affect seafood distributors and users are mentioned in the following paragraphs.

One of the cornerstones of the system is tagging shellfish to indicate where and when it was harvested. In the event of a problem, it is then possible to track it back to the source and to recall any other shellfish that may have come from the affected area. Shellfish tags include the name and license number of the original shipper and show when and where the product was harvested. When shellfish are removed from their original containers for sale or consumption, the wholesaler, retailer or restaurant must retain the tag for at least 90 days. In practice, the best advice is that you retain the tags indefinitely: you never know how long it may take someone to decide to sue because they ate something that disagreed with them and have finally decided to blame the shellfish they bought. A typical tag is shown in Figure 4.1. The rules apply equally to imported shellfish.

Any containers of live clams, mussels or oysters delivered without tags should be firmly rejected. It is possible that the tag fell off: that is the supplier's problem. Don't let it be yours.

Live shellfish should be kept cool but should not be in direct contact with melting ice. Ice is required to maintain temperature and moisture, but the product should be on pallets so that it does not sit in fresh water (which kills these animals). Cool and moist conditions are essential for their survival. Check that arriving product appears to have been stored and shipped in suitable conditions. Figures 4.2 and 4.3 show some standard containers used for shellfish. Keep all shellfish containers off the floor and away from melting ice. Most live molluscs are sold in volume containers, usually related to bushels. Most states have regulations defining their particular bushel, bag or other unit of measurement. The weight of a bushel of oysters will vary substantially, partly with the season and partly from area to area as oysters from different places may have different ratios of shell to meat. Weight is totally unreliable as a guide to the volume received.

Oysters, clams and mussels should be free of barnacles, seaweed

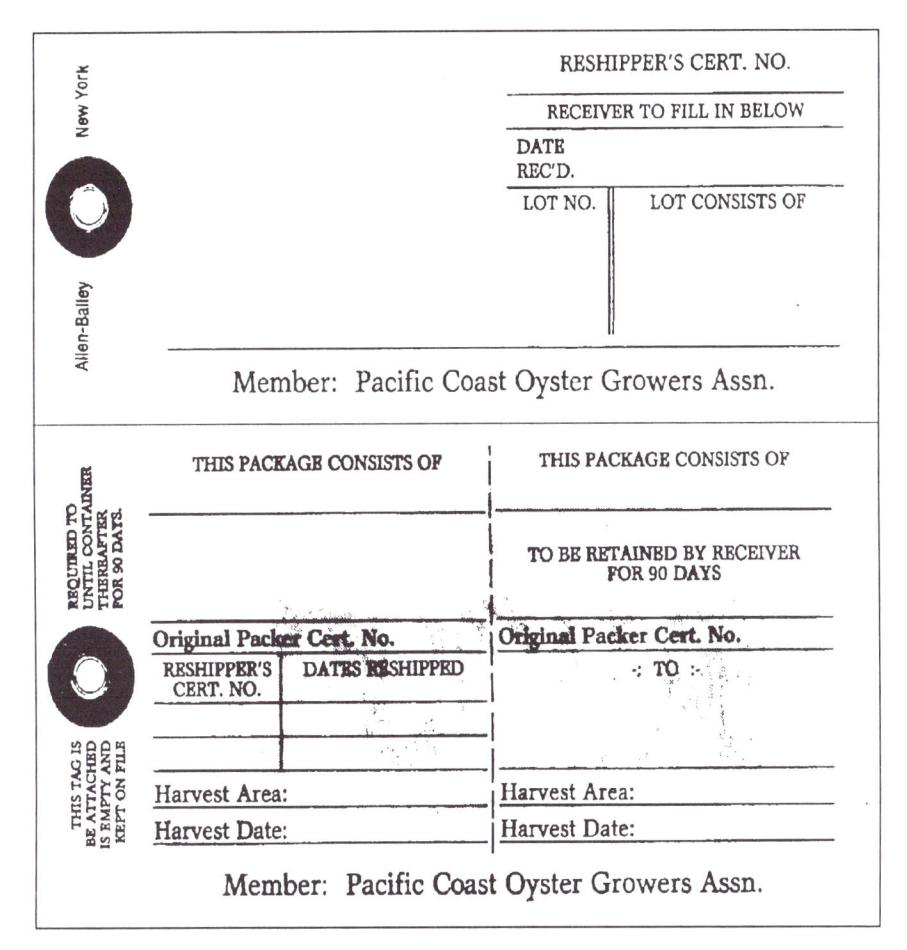

Figure 4.1. Tag for container of live shellfish.

and other unwanted detritus. They should be firmly closed. Dead shellfish usually gape open. If an open oyster or clam closes when it is tapped on a table, it is still alive and usable. Do not use dead shellfish. Check animals at the bottom of sacks to make sure that they have not been thrown around and cracked. Cracked shellfish do not last long and in any case should not be used as they are liable to pick up bacterial contamination.

If you store or display molluscs in live holding tanks, make sure that the manufacturer's instructions for the tanks are followed. It is quite easy for shellfish in tanks to grow large colonies of potentially harmful bacteria. Texas banned the sale of molluscs from live tanks in 1990 because of possible risks to human health.

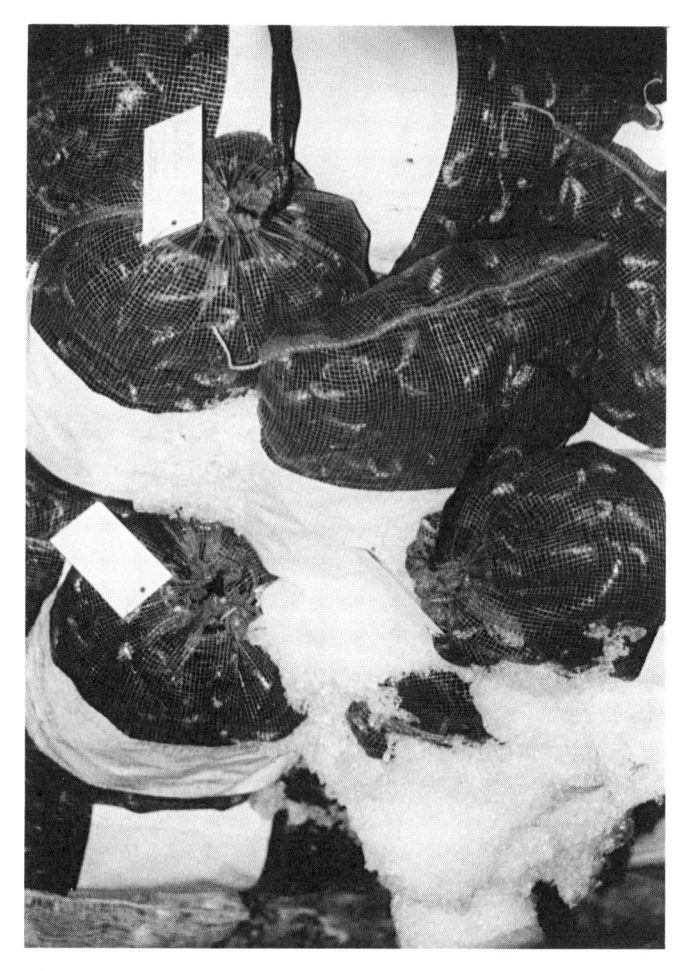

Figure 4.2. Onion bags are frequently used for live shellfish.

Oysters

Live oysters are sold as single shellfish or in clumps, which are cheaper. If you have bought clumps, make sure that there is not an excessive amount of old, dead shell: you are paying for live oysters. Size grading tends to be eclectic, but if your supplier specified a size, check that you got what you ordered. Pacific oysters (*Crassostrea gigas*) grow much larger than the eastern oyster (*Crassostrea virginica*) and can be a problem for half-shell use in areas accustomed to the smaller oyster. The two species look very similar.

Figure 4.3. Bushel boxes provide better protection for fragile soft-shell clams.

Clams

Hard shell clams, which are mostly *Mercenaria mercenaria*, the quahog, cherrystone and littleneck of the East Coast, are quite tough and will withstand more abuse than oysters and mussels. They are, however, brittle and the shells crack if bags are thrown or dropped. Check clams from the bottom of sacks for signs of damage. Soft-shell clams *(Mya arenaria)* are extremely delicate. These are usually packed in baskets (Figure 4.3). The shells do not completely enclose the animal, which is very susceptible to death from drying out. Dead soft-shell clams have limp, extended necks.

Mussels

Mussels should be well scrubbed and all barnacles and other growths removed. Beards, the black byssus threads which the mussels use to attach themselves to their growing place, may be trimmed off to improve appearance but should not be pulled out until the mussels are about to be cooked. Mussels benefit from being packed tightly into bags or sacks. This helps to prevent them from opening and drying out. Again, check the mussels at the bottom of the sack for signs of damage from rough handling.

Scallops

These are also filter feeding molluscs. In the past, they were not subject to the NSSP because normally only the adductor muscle is eaten and this does not usually contain bacteria from growing waters or concentrate harmful marine toxins. In recent years, some distributors have been offering whole scallops for steaming or for serving, European style, with roe. This practice is risky and not recommended, because the roe and viscera of scallops do concentrate marine toxins. In July 1990, scallops sold live, whole or with roe became subject to the provisions of the NSSP. It is likely that all scallop meats will be included in the Program soon, which means that harvesting areas will be inspected and all processors and dealers will have to hold state licenses.

If you buy live scallops, remember that they are extremely delicate and have very short lives out of their natural environment. If possible, ascertain that your supplies have come from an area which is approved for other shellfish harvesting. This will help to ensure that the scallops have not picked up red tide or related toxins. Check that the scallops are alive and do not accept or use any that are dead.

Whelks

The East Coast dog whelk (scungili) is sometimes offered whole and live. This is a meat-eating shellfish and only the mantle and foot are eaten. Whelks are not subject to tagging rules and do not need to be. They are extremely sturdy animals and can survive for months out of water if kept cool and damp. Nevertheless, like any other shellfish, they can be cracked or broken by careless handling. Check that the whelks are alive and do not accept or use dead ones.

Shucked Molluscs

Scallops are shipped in cloth bags or in plastic pails. They build up considerable odor in the container. Do not attempt to judge the

Figure 2.2 Though still colorful, the dead mahimahi loses much of its brightness and iridescence.

Figure 2.3 American red snapper *(Lutjanus campechanus)* retains its color and appearance well.

Figure 2.4 The gills of the lower fish are turning brown, indicating it is a little older than the other fish, though both are still very fresh.

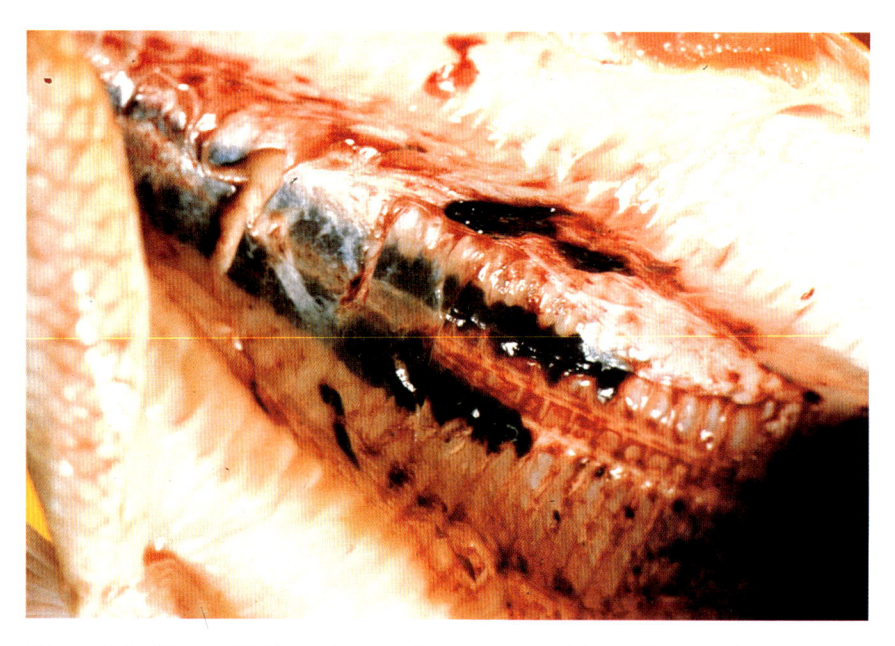

Figure 2.5 The red kidney has not been removed from the membrane at the top of the belly cavity.

Figure 2.6 A cleanly eviscerated fish.

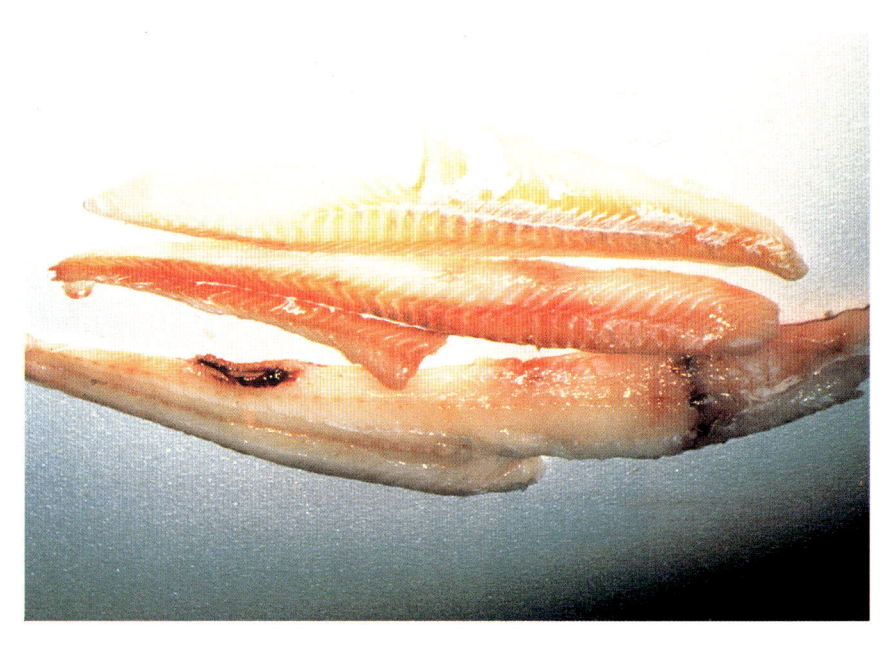

Figure 2.18 Bruising, blood spots, ragged edges and remnants of skin are among the visible defects of these fillets.

Figure 2.19 The subcutaneous layer on this fillet is silver and shiny, which looks fresh and appealing on display.

Figure 2.20 The subcutaneous layer is oxidizing and turning unattractively brown.

Figure 3.10 Freezer burn.

Figure 3.11 Rancidity.

Figure 3.12 **Frost on breaded product.**

Figure 3.13 **Coating defects.**

Figure 4.4 **Oyster liquor should be clear, not opaque.**

Figure 4.5 **Amounts of oyster liquor vary.**

Figure 5.2 Abalone steak (the missing piece was used for electrophoresis to determine the species).

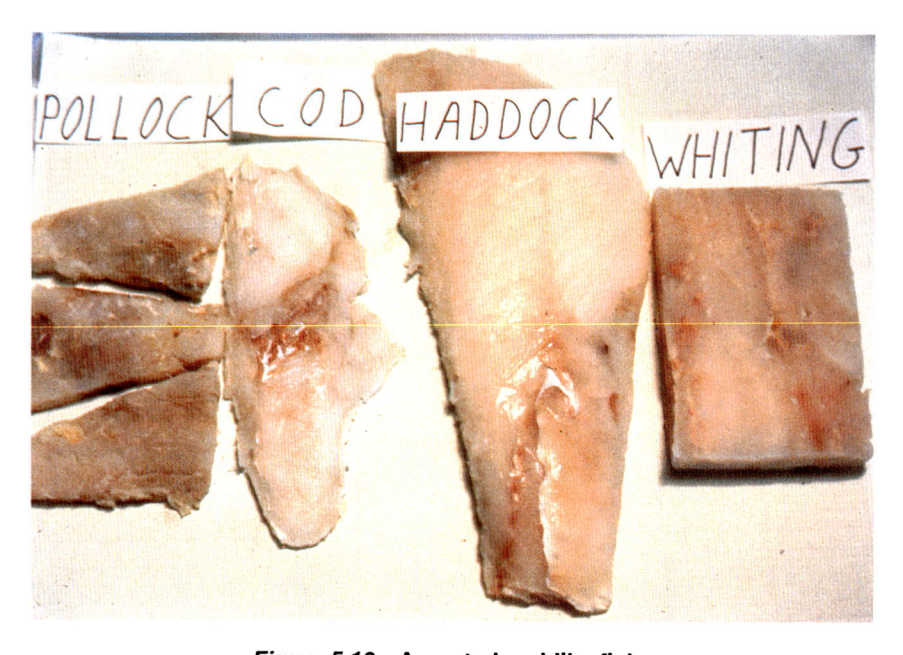

Figure 5.10 Assorted cod-like fish.

Figure 5.12 Florida stone crab claw (top) is more clearly marked than the South American claw.

Figure 5.21 Mahi-mahi is also known, unfortunately and confusingly, as dolphin.

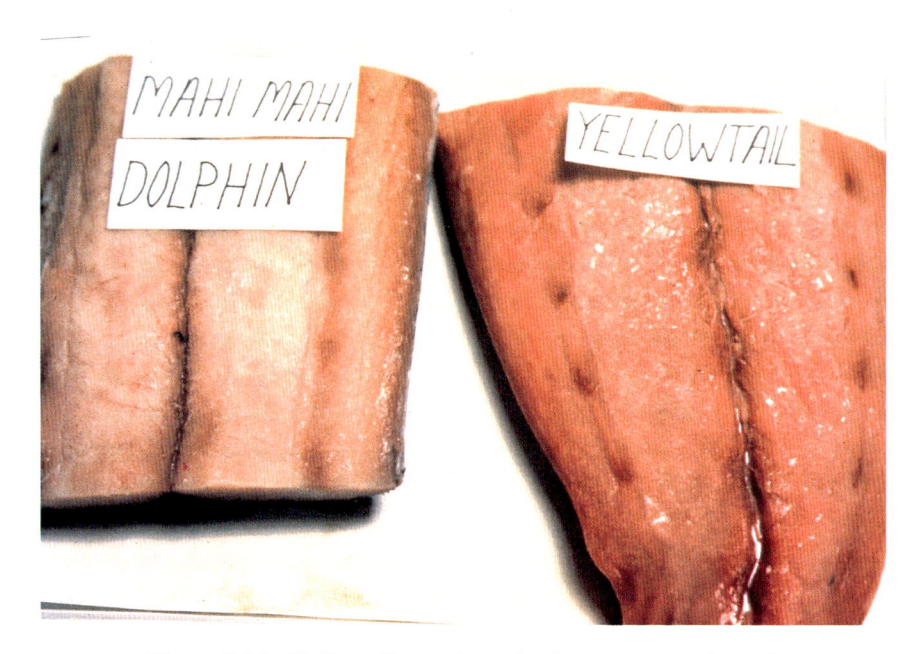

Figure 5.22 **Yellowtail may be substituted for mahi-mahi.**

Figure 5.23 **Lake perch.**

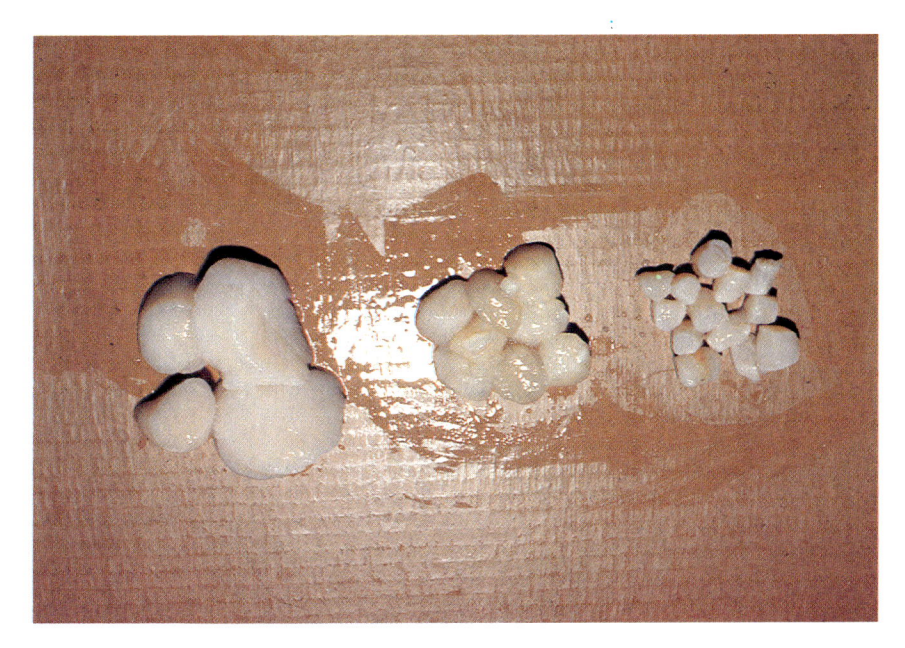

Figure 5.25 Left to right: Sea scallops, bay scallops and calico scallops.

Figure 5.26 A genuine American red snapper *(Lutjanus campechanus)*.

Figure 5.27 Six snappers: Left to right, top to bottom: Silk, gray, vermilion, queen, red, mutton.

Figure 5.28 Top: Spotted rose snapper. Below: American red snapper.

Figure 5.29 Pacific ocean perch *(Sebastes alutus)*. Photo: Donald Kramer, University of Alaska.

Figure 5.30 Widow rockfish *(Sebastes entomelas)*. Photo: Donald Kramer, University of Alaska.

Figure 5.31 **Yelloweye rockfish *(Sebastes ruberrimus)*. Photo: Donald Kramer, University of Alaska.**

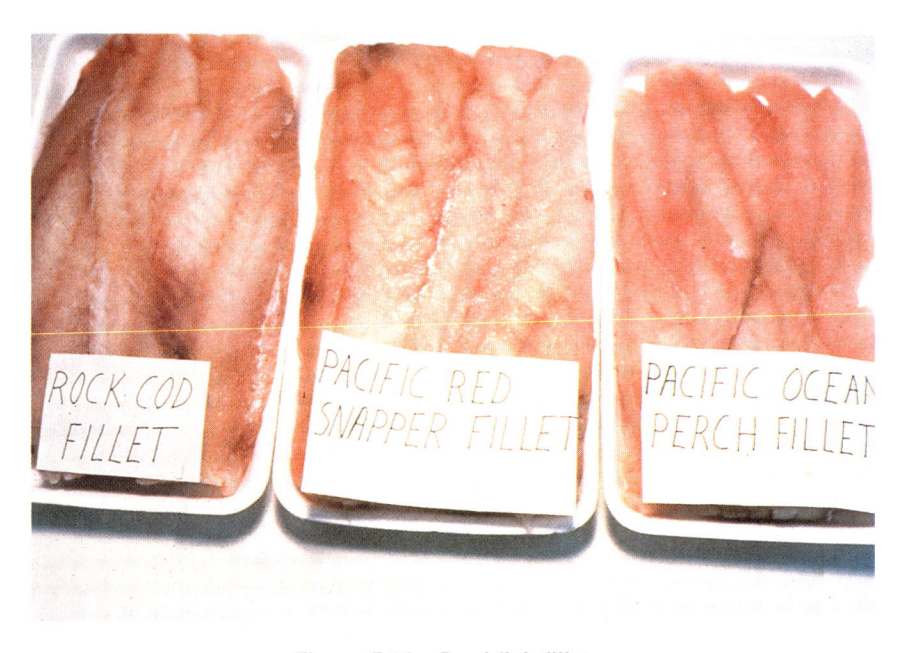

Figure 5.32 Rockfish fillets.

Figure 5.33 Rockfish fillets, skinned side up.

Figure 5.35 Swordfish (on right) and shark.

Figure 5.36 **Shark on top of swordfish.**

Figure 6.7 **Six sizes of shrimp, from 21/25 down.**

freshness of scallops until they have been rinsed or until the odor has been allowed to dissipate after the containers have been opened. Scallops, especially if frozen, are sometimes dipped in phosphate solutions which help to reduce drip loss. Excessive dipping enables the muscle to absorb extra water. Foaming and a soapy feel to the product indicate phosphate abuse.

Shucked oysters, mussels and clams must be in containers marked with the packer's license number and the date of packing. Containers without this information should always be rejected. They are illegal as well as potentially dangerous. Oyster liquor should be clear, not opaque (Figure 4.4, see color insert). Cloudiness is often a sign that the oysters are aging. Liquor amounts vary (Figure 4.5, see color insert). Determining net weights of shucked shellfish, fresh or frozen, is tricky. See Chapter Seven for suggested methods.

Mollusc meats must be well iced. Temperature control is particularly important or bacteria can multiply fast and become a hazard.

Live Lobsters

Northern lobsters (*Homarus americanus*) are almost always sold live. They must be kept cool and damp while out of water. Make sure they are alive when they reach you. Lobsters are subject to federal minimum size regulations. (States may determine other sizes for lobsters caught within their three mile territorial seas and sold within the state. Check with the fisheries department.) At the beginning of 1991, the federal minimum size increased to $3\frac{9}{32}$ inches. From the beginning of 1992, the size is $3\frac{5}{16}$ inches. Size is defined as the length of the carapace from the back of the eye socket to the back of the shell covering the body (see Figure 4.6). The penalties for possessing undersize lobsters can be severe. Under the federal Lacey Act, these penalties extend even to states which have no lobster supplies or regulations of their own.

It is also illegal in the United States to possess berried lobsters (females carrying eggs externally) (see Figure 4.7).

Lobsters should always be shipped right side up. They die much sooner if they are upside down. Cool temperatures discourage them from moving. Claws should always be banded or pegged so that lobsters cannot damage one another.

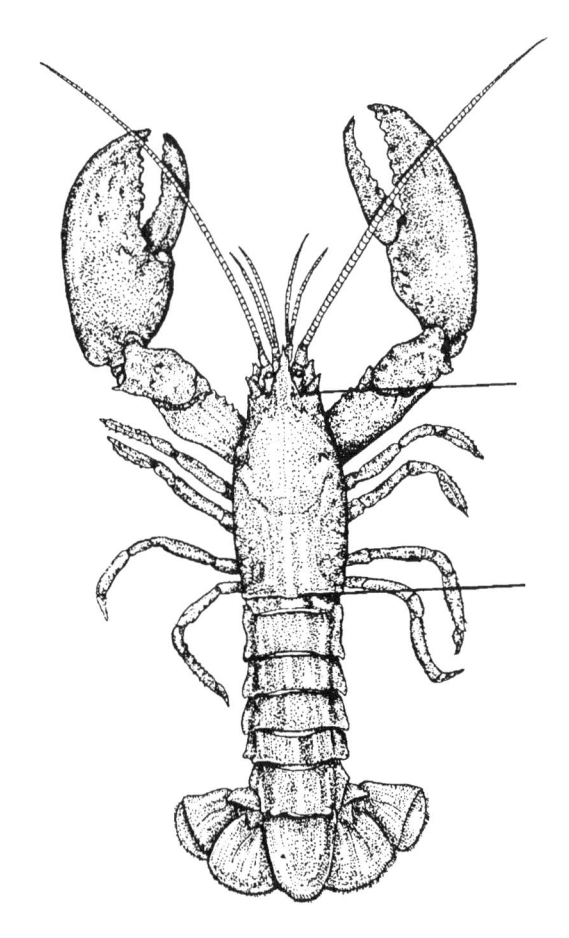

Figure 4.6. Lobsters are measured from the back of the eye socket to the back of the carapace.

If you put lobsters into live tanks, never mix them with molluscs such as clams, as each can contaminate the water in ways harmful to the other.

Crabs

Blue crabs provide hard crabs, softshell crabs and crabmeat. The species, *Callinectes sapidus*, is the most important commercial crab species in the United States. Both male and female crabs are used (Figures 4.8 and 4.9). Meat products vary greatly in specification. The larger pieces from the body of the crab are the most expensive.

Crabs 51

Figure 4.7. A berried lobster is a female carrying its eggs externally. Possession of berried lobsters is illegal in the United States.

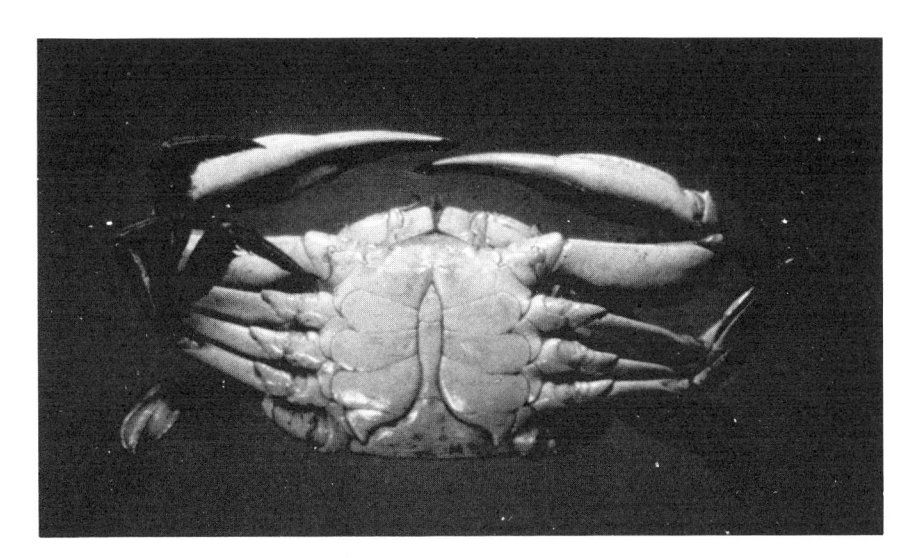

Figure 4.8. Male blue crab.

52 Quality and Freshness—Shellfish

Figure 4.9. Female blue crab

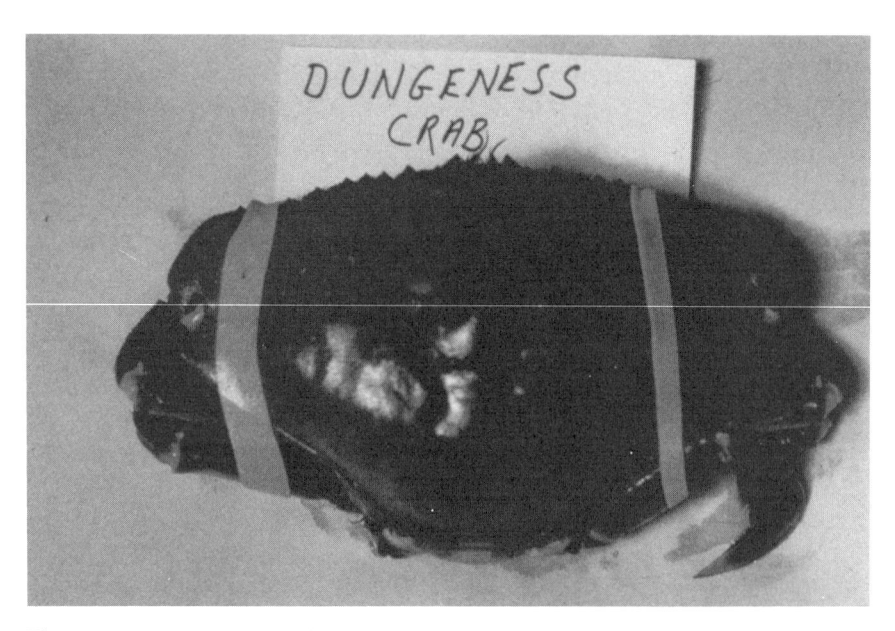

Figure 4.10. Dungeness crabs are measured across the widest part of the top shell.

Figure 4.11. Snow crab. Bairdii (top) is larger and more spiny than opilio.

For blue crab specifications, see *Product Quality Code* from the Southeastern Fisheries Association. Packs should contain the type of meat stated and not meat from a less desirable part of the crab.

Live crabs should be packed tightly in baskets to reduce movement and prevent the crabs from damaging each other. They should be kept cool and used quickly.

Dungeness crabs are subject to minimum size controls, measured as indicated in Figure 4.10. Female crabs may not be taken. Dungeness crabs are usually cooked at sea. Check for any signs of decomposition and accept only fresh product.

There are two types of snow crab: bairdii and opilio (Figure 4.11). Opilio is smaller and less valuable than bairdii. It is smoother and easy to distinguish. Snow and king crab are size graded and the sizes should be checked from time to time. Larger sections are worth more than smaller ones.

Chapter 5

Substitutions

What the federal government likes to call "economic fraud" has three main elements in the seafood business. The most complex to deal with is substitution of an inferior species for a more valuable one. The other two, supplying inferior size or quality and supplying short weight, are covered in the next chapter.

Seafood species substitution is possible partly because of the large numbers of different fish and shellfish that are used in the United States. Some observers consider that there are at least 300 species commonly sold. Many more species are used commercially from time to time: New York's Fulton Wholesale Fish Market has seen almost 1,000 different species pass through in the course of some three years. It is difficult, if not impossible, for anyone to be sufficiently familiar with all of these to be able to tell at a glance what the species might be. In practice, most of the seafood we use comes filleted or prepared in some way, so that the major features which aid identification, such as the heads, fins and scales, are missing.

Add to the great variety of seafoods the similarity between some of them and the problems are greatly compounded. Many soles and flounders look identical once they have been filleted. So do many of the rockfishes. King salmon and coho salmon are easily distinguishable by the color of the gums; once the head is removed, the identification is much more difficult. Without the fins and scales, it is virtually impossible in most circumstances. If price differentials are great, dishonest suppliers are tempted to label a less expensive fish as a more valuable one.

There are laboratory tests which can identify the flesh of many raw fish and shellfish, but these tests are expensive and time consuming. In a situation which might lead to a lawsuit, such tests are a requirement. In daily business, the best defense is to be aware of possible problems, check everything carefully and use reliable suppliers. Many times, it is the bargain hunter who gets caught. To repeat a cliche, if a deal seems too good to be true, it probably is.

Some of the price differentials which attract dishonest practices make little sense. If Chinese white shrimp is so similar to domestic white shrimp (Figure 5.1) that it can be repacked and passed off as U.S. product, why does it sell at a substantial discount? Mako shark used to be sold as swordfish until buyers realized that it was a good product in its own right and started buying it as shark. This raised the price and made substitutions less profitable.

This chapter contains information on some of the more frequent substitutions. Bear in mind that seafood is a dynamic and developing industry. Just as new products are launched every day, new scams are developed and perpetrated. As prices change, so do the possibilities of substitutions. The most vulnerable products are the

Figure 5.1. Left to right: 31/35 domestic white; 26/30 Chinese white; 21/25 Chinese white.

more expensive ones and those that become scarce. If you are flexible enough to be able to change what fish and shellfish you use as prices and availabilities change, you are less likely to be the victim of this type of fraud.

Abalone

Abalone is a very expensive shellfish and supply is limited. There are seven species found in Pacific waters of the United States, all members of the genus *Haliotis*. In the United States, most of the consumption was in California, but the establishment of Japanese restaurants has created demand throughout the country. Cuttlefish and giant squid have both been substituted for abalone steaks. Figure 5.2 (see color insert) shows an abalone steak. It has no membrane. One cuttlefish steak in Figure 5.3 was processed with a meat tenderizer, which perforates the flesh with needles. The other steak still has the membrane intact and will curl when it is cooked. The presence of the membrane or the needle holes from the tenderizing machine are sure indicators that either cuttlefish or squid are masquerading as abalone.

There is a markedly inferior shellfish, usually called loco in its native South America, which has been sold in the United States as canned abalone. This is totally unlike abalone in appearance, taste and texture. Cans of loco labeled as abalone are liable to seizure by Food and Drug enforcement officials.

Figure 5.3. Cuttlefish steaks imitating abalone.

Clams

Littleneck clams are the smallest size description of the eastern hard shell clam, *Mercenaria mercenaria*. They are the most expensive size. They are mostly served raw on the half shell. Mahogany clams, also called ocean quahogs and black clams, are sometimes substituted. The ocean clam is darker and has a dark brown, hairy layer on the shell. Mahogany clams are tougher than genuine littlenecks and the meat is darker. They are mostly used for clam strips and minced clam products.

On the Pacific coast, there is a littleneck clam *(Protothaca staminea)* which is similar in size to the eastern littleneck. This clam is too tough to be used raw and is sold for steaming. Although it is unlikely that Pacific littlenecks would be sold deliberately to markets expecting eastern clams, there is always the possibility of confusion. For illustrations of all the major clam species used in the United States, see this author's *Shellfish—A Guide to Oysters, Mussels, Clams, Scallops and Similar Products for the Commercial User*, published by Van Nostrand Reinhold/Osprey Books.

Cod, Haddock and Similar Species

Cod remains a staple American seafood. It is related and very similar to haddock and closely related to pollock, hake and whiting. Atlantic cod and Pacific cod are a little different. Atlantic pollock and Pacific pollock are substantially different. To add to the complications, the relative prices for all these fish fluctuate so that the incentives to substitute one for another change at different times.

Probably the most common substitutions are cod for haddock and Pacific pollock for both cod and haddock. Haddock is easy to identify if it has the skin. Figure 5.4 shows haddock. Note the black lateral line, which is the major distinguishing mark. The lateral line on cod and pollock is white. Haddock also have a "thumbprint" on the skin near the nape, but this is not always easy to see, especially if the scales have been removed. Haddock should always be bought with the skin on. Skinless haddock, if it costs more than skinless cod, is a waste of money for most buyers. Few if any people can distinguish between the tastes of the two species.

Figures 5.5 and 5.6 show skinned fillets of Atlantic and Pacific

Cod, Haddock and Similar Species 59

Figure 5.4. Haddock. Note the distinguishing black lateral line.

Figure 5.5. Atlantic cod, skinned, showing the silvery subcutaneous layer.

Figure 5.6. Pacific cod, skinned.

cod. The clearest distinction is the silvery subcutaneous layer on the Atlantic fish. The Pacific cod is often described as "true cod," though the FDA does not approve of the term. In both foodservice and retail use, the two species are largely interchangeable.

Atlantic pollock (Figure 5.7) and Pacific pollock are sometimes substituted for both cod and haddock. Cusk (Figure 5.8) is another gadoid (cod-like fish) which looks very similar to cod and may be substituted for cod occasionally.

Scrod is not a fish: it is a size description for small cod, haddock and pollock. Any fish described as scrod should also have the species name attached to it. Scrod cod is small cod; scrod haddock is small haddock. If you buy "scrod," there is no telling what you might be shipped and no way you can argue with your supplier about it. Scrod continues to be a popular fish in the United States. Nevertheless, it does not exist.

Whiting is an inexpensive fish, usually processed into blocks for the manufacture of fish portions. Deep-skinned (which means the subcutaneous fatty tissue is removed), it is impossible to tell whether the fish is whiting or Pacific pollock without laboratory testing such as iso-electric focussing. Figure 5.9 shows a whiting fillet.

All of these gadoid fish are similar: Figure 5.10 (see color insert) shows a selection of them.

Cod, Haddock and Similar Species 61

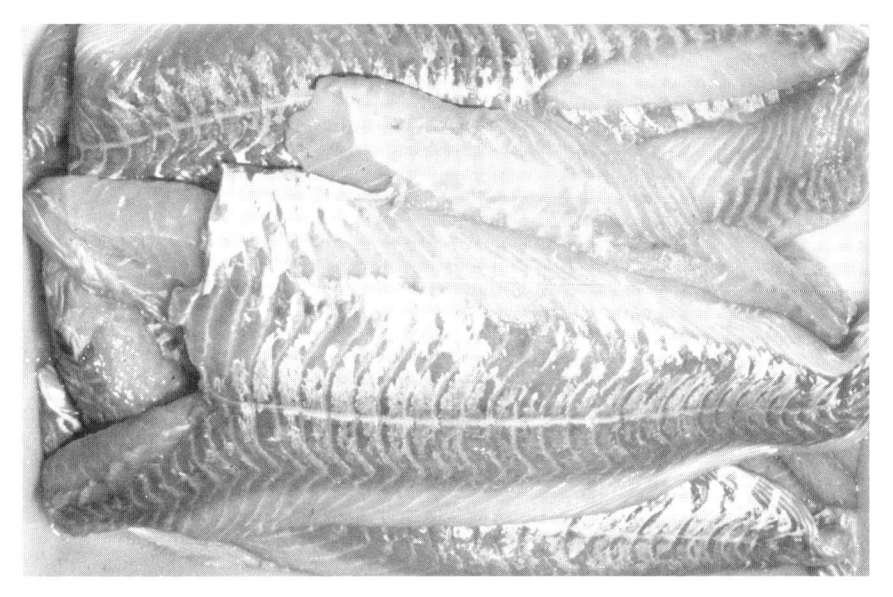

Figure 5.7. Atlantic pollock has grayish meat, which is quite white when cooked.

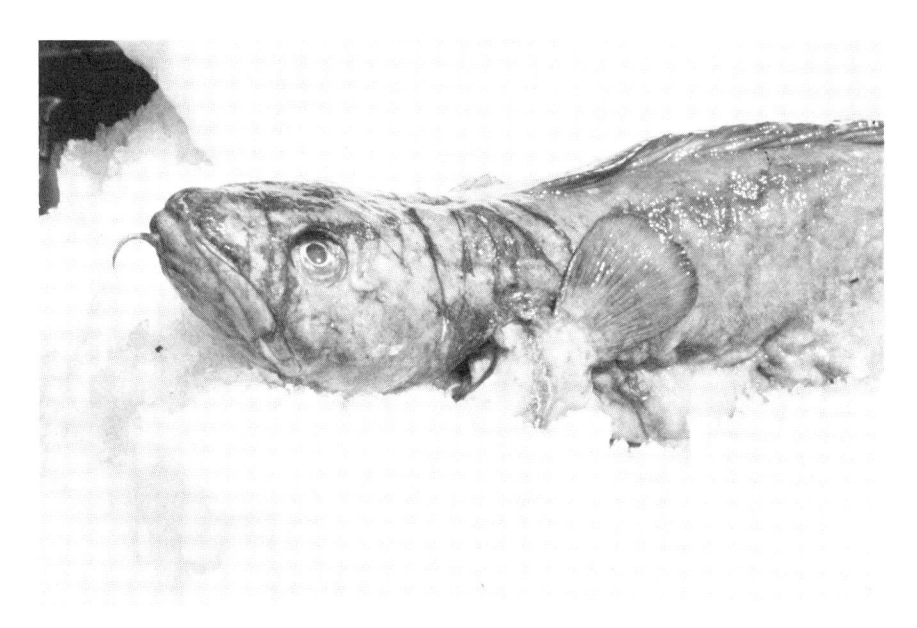

Figure 5.8. Cusk is a large gadoid sometimes used instead of cod.

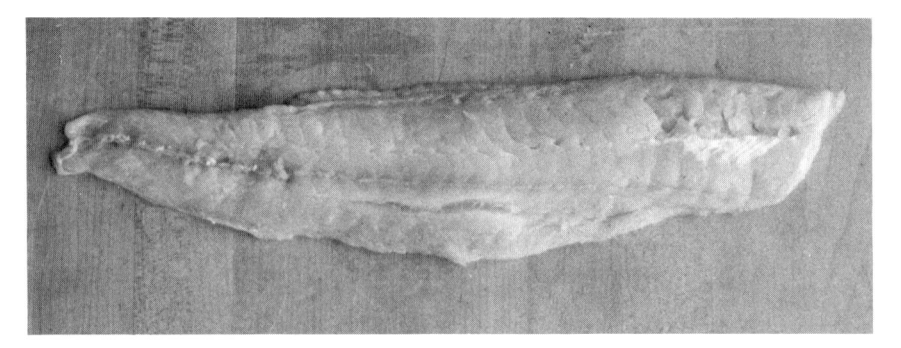

Figure 5.9. Whiting is smaller and has softer meat than most other gadoids.

Crab

King crab is the most expensive crab normally found in commerce. There are three species, all of which come from the North Pacific:

Paralithodes camtschatica . . . . Red king crab
Paralithodes platypus . . . . Blue king crab
Paralithodes brevipes . . . . Brown (or deepwater) king crab

King crab meat is an extremely expensive product. It is usually described as "60/40," meaning that 60 percent of the block is body meat and 40 percent is leg meat. It is usually packed in three layers. The bottom layer, usually 25 percent of the whole pack, should be merus meat, which is pieces taken from the largest segment of each leg. The center layer is the white body meat (it should not include meat from the tail or from the tips of the walking legs) and the top layer, which is about 20 percent of the block, is red meat from the legs. Packers try to give about 55 percent body meat, increasing the leg meat proportion over the specified proportion to 45 from 40 percent.

King crabs are caught by fishermen from the United States, Japan and the USSR. There is a smaller, related species, *Lithodes antarctica*, which is imported from Chile and Argentina. This may be illegally labeled as "Alaskan" crab or "king" crab. The species is smaller than the North Pacific king crabs and the standard of packing is generally inferior. However, because of its price advantage, the South American product is now widely used in the United States.

Jonah and rock crabs are small east coast crabs. The Jonah crab

(Cancer borealis) can be distinguished from the rock crab *(Cancer irroratus)* by its thicker legs, scalloped shell and black tips on the claws. Meat from the rock crab is being sold as an inexpensive substitute for west coast Dungeness crab *(Cancer magister)* meat. Figure 5.11 shows rock crab meat.

Stone crab claws are a Florida specialty, fished only on the Gulf Coast. Fishermen remove the black-tipped claws and return the crab to the sea to regenerate more claws. South American stone crabs are sometimes offered as a substitute. These have less shiny shells and fewer markings than the Florida claws. Figure 5.12 (see color insert) shows the comparison.

Kamaboko, imitation crabmeat, is too often substituted for crabmeat at restaurant and retail level. This product is made from surimi, which is a washed, minced fish flesh product from which taste and texture have been removed. Surimi is used as the raw material for a wide range of imitation seafoods and is even being made into imitation meat products, such as hot dogs. Kamaboko can be made to look very much like crabmeat (Figure 5.13). It is a perfectly good product, but its fraudulent use as a substitute for much more expensive crabmeat harms the reputation and long term viability of everyone in the seafood business.

Figure 5.11. Rock crab and meat.

Figure 5.13. Imitation crabmeat.

Flounders and Soles

The distinction between flounder and sole is not clear. According to the National Marine Fisheries Service, a *Pseudopleuronectes americanus* under 3½ pounds is a blackback flounder, while the same species over 3½ pounds is a lemon sole. The Food and Drug Administration, in its 1988 *Fish List*, defines the same fish as either winter flounder or lemon sole. All the legal flounder and sole names specified in the *Fish List* are shown in the Appendix in Table A.2. Note that the FDA is the arbiter of seafood nomenclature. The names they require should always be applied.

The confusion is partly because consumers in some regions prefer the word sole while others prefer the word flounder. Since the fish is very similar, or even, as mentioned above, may be the same, sorting out the names further than is required to conform to the law seems pointless.

Some flounders and soles are, however, worth much more than others. Gray sole *(Glyptocephalus cynoglossus)*, also called witch and witch flounder, is a premium fish. It has long, slender fillets (see Figure 5.14) which are quite distinctive. Winter flounder, still most commonly known as blackback, despite the FDA's disapproval, is the commonest East Coast flatfish. Fillets of greenland turbot *(Reinhardtius hippoglossoides)* have been frequently substituted.

Figure 5.14. Gray sole (on left) is longer and narrower than winter flounder (on right).

Often, the turbot is whiter and looks more attractive. Although it is perfectly palatable, it is inferior in taste and texture and does not hold up under some cooking methods. Turbot is larger than flounder, with fillets as large as two pounds not uncommon. Only the smaller fillets can be reasonably passed off as flounder.

Arrowtooth flounder *(Atheresthes stomias)* is similar in size and appearance to greenland turbot. Arrowtooth contains an enzyme which tends to break up the flesh when the fillet is cooked. Arrowtooth is consequently a very cheap and poorly regarded fish. If arrowtooth is substituted for turbot and the turbot is then substituted for flounder, this double fraud could be extremely costly. Canadian greenland turbot from the Atlantic coast is generally regarded as superior to the Pacific product purchased from Japan and South Korea. Canada also has stocks of arrowtooth flounder on their west coast. This fish is legally labeled as turbot in Canada. Make sure that you do not get caught by this confusion. Scientists in Alaska, which also has substantial arrowtooth resources, are working to develop ways to process and package arrowtooth so that it does not collapse when cooked.

Greenland turbot must not be confused with "genuine" European turbot, *Psetta maxima* or *Scophthalmus maximus*, which is a large, high quality flatfish that is usually imported headless and dressed and served by fine restaurants in steaks rather than fillets. This fish is now being farmed in Europe and may become more readily available. However, if people are served greenland turbot instead, it is unlikely that a market will develop for the real thing.

Turbot is not the only flatfish with a name designed to confuse. The European Dover sole *(Solea vulgaris)* is the original sole and is at least as expensive as genuine turbot. In the United States, it is generally available scaled and dressed, not filleted. Do not confuse this fish with Pacific dover sole *(Microstomus pacificus)*, which is a soft flatfish, once called slime sole, giving a small fillet. Californian dover sole is usually unacceptably soft, but the Alaskan fish is reportedly better. Do not confuse the Pacific species with the European fish.

One simple technique which may sometimes aid identification is to determine if a flatfish is right-eyed or left-eyed. Hold a fish with its belly towards you and the eyes on top. If the head is at your left, it is a left-eyed flatfish. If you have a pair of fillets, you can reconstruct the fish sufficiently to be able to tell if the eyes were on the left or the right (see Figure 5.15). Fluke (summer flounder) is left-eyed, while winter flounder is a right-eyed fish. For information on which species is which way around, see *The New Frozen Seafood Handbook*.

Halibut

Most halibut sold in the United States comes from the Pacific and is *Hippoglossus stenolepsis*. Atlantic halibut *(Hippoglossus hippoglossus)* is now comparatively rare. The small catches are mostly sold fresh in the Northeast. California halibut *(Paralychthys californicus)* is a smaller fish with less solid flesh. It is not usually marketed much outside California. Halibut has firm, very white meat and superb flavor. Fresh or frozen, it is one of the best fish to eat. Lingcod *(Ophiodon elongatus)* is occasionally substituted, even though it looks remarkably different as a whole fish. Figure 5.16 shows lingcod and halibut steaks together. Note that the lingcod has more red (dark) meat than the halibut. The skin (Figure 5.17) is a clear indication. The skin of the lingcod has spots, which the halibut lacks.

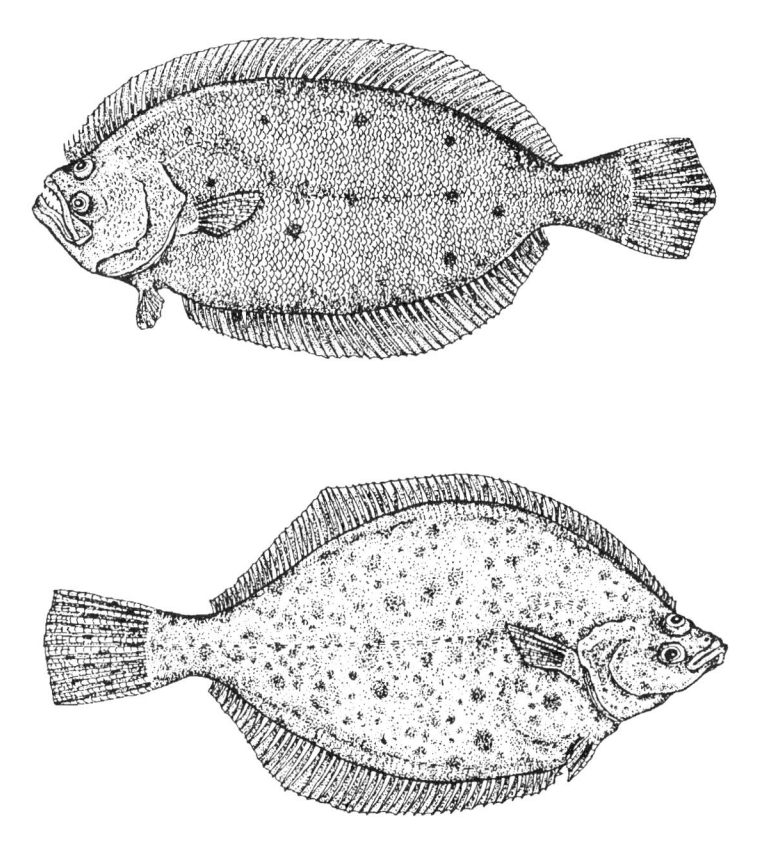

Figure 5.15. Left-eyed flatfish at top, right-eyed flatfish underneath. You can reassemble fillets to indicate which way the head was. This can help in identifying the fish.

Lobsters

Northern Lobster

The American lobster (Figure 5.18) has two large claws. Although it is mostly sold live, weak or damaged lobsters may be cooked and offered whole, as picked meat or as cooked tails. In some states, possession of tails is against the law, to help enforce minimum size regulations. Tails of northern lobsters can be distinguished from those of rock (spiny) and slipper lobsters by the tail fan or fin. In the northern lobster, this is hard and opaque. Spiny and slipper lobsters have leathery, rather pliable tail fans which are somewhat translucent at the rear (Figure 5.19).

Langostino meat is substituted for lobster meat, but probably

68 Substitutions

Figure 5.16. Lingcod and halibut steaks.

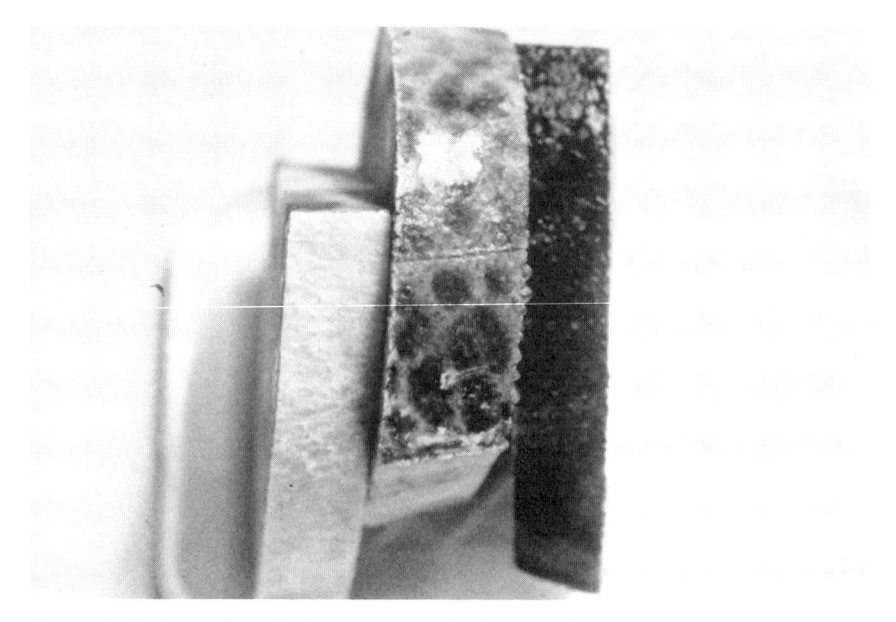

Figure 5.17. Lincod and halibut steaks. The lingcod has distinctively spotted skin.

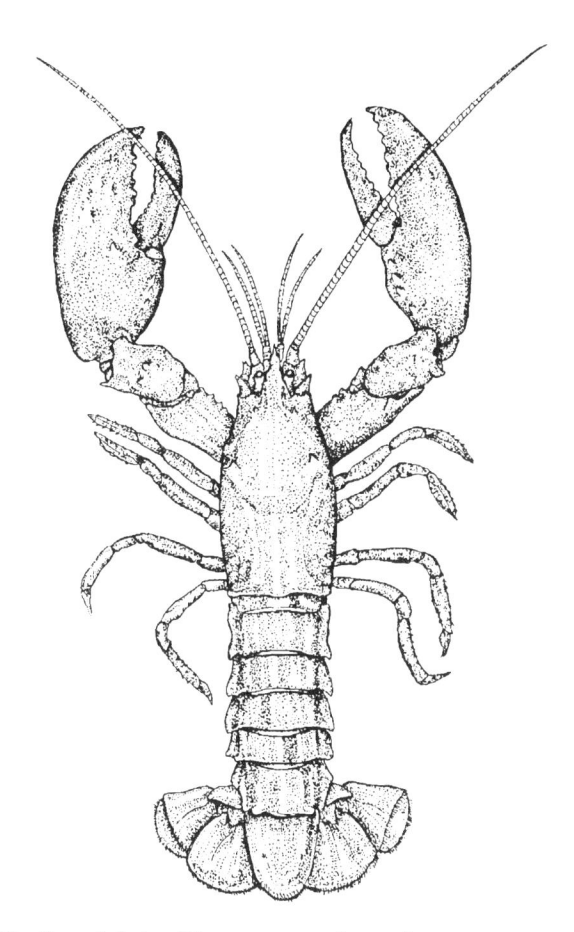

Figure 5.18. Northern lobster (Homarus americanus).

only at the final level of the trade, where sandwiches are made. Langostinos have small meats resembling cooked and peeled shrimp. Lobster meat contains all the different parts of the lobster and has large fibers and chunks.

Spiny or Rock Lobsters

Spiny or rock lobsters (the two terms are completely interchangeable) are differentiated in the industry by their origin from cold or warm water. Although this distinction is not clear (it seems to be based more on perceptions of water temperature in the countries of origin rather than on any precise evaluation of the species'

70 Substitutions

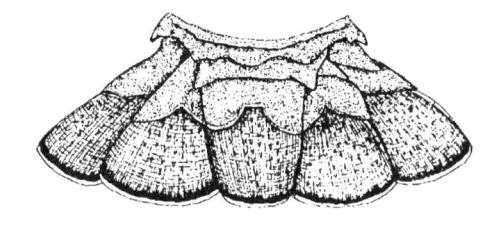

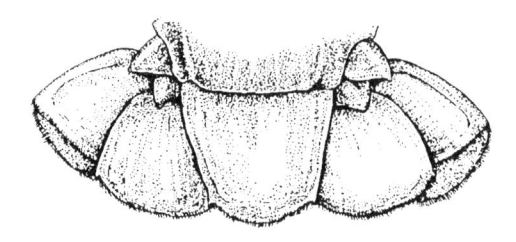

Figure 5.19. Tail fans of spiny lobster (top) and clawed lobster (below).

preferred habitat), it has considerable effects on the price. If you buy cold water lobster tails, such as those from New Zealand and Australia, you will pay substantially more for them and need to know that you are getting the right product.

Identifying rock lobsters from the tails alone can now be done, thanks to the work of Dr. Austin Williams at the National Systematics Laboratory. Complete details of how to identify the commercial species found in the United States are given in *Lobsters of the World—An Illustrated Guide*, published by Van Nostrand Reinhold/Osprey Books. If you have doubts that your suppliers are sending you the right sort of lobster tails, you should study the distinctions discussed in that book. Note that color is not a very reliable guide: the color of individual lobsters of the same species can vary greatly.

Slipper lobsters (flat lobsters) are an inferior product group. There are a few species which produce tails comparable in size with rock lobsters, but are worth much less in the marketplace. The same resource mentioned above will enable you to identify the difference between rock and slipper lobsters. Basically, the distinction is that rock lobsters have spines on the end of the side plates

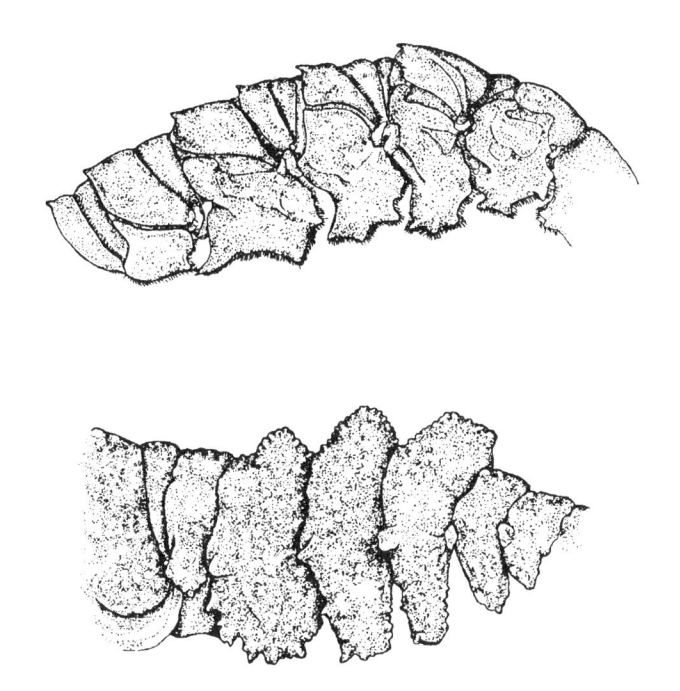

Figure 5.20. Rock lobster tail (top) compared with slipper lobster (below).

covering the tail; slipper lobsters have irregularly shaped edges to the plates, often ending in tubercles or knobs. The tail segments of slipper lobsters are also noticeably flattened. See Figure 5.20 to compare rock and slipper lobster tails.

Lobster tails are particularly subject to over-glazing and short weight. See the next chapter for more details.

Mahimahi

Mahimahi is the Hawaiian name for *Coryphaena hippurus*. The fish is also known as "dolphin-fish." This term should be avoided because of the obvious confusion with the marine mammal dolphin, which it in no way resembles. Mahimahi is a fish, not a mammal. It is easy to recognize by the yellowish skin and the two lines of dark meat in the flesh (Figure 5.21, see color insert). When first caught, the skin of a mahi is very colorful, but the colors soon fade, leaving only the yellowish streaks (see color Figure 2.2). California yellowtail *(Seriola lalandei)* a member of the jack family, provides a similar looking fillet. The flesh looks very similar (Figure 5.22, see color insert), but the skin on a yellowtail has yellowish spots rather than streaks.

Orange Roughy

Orange roughy *(Hoplostethus atlanticus)* comes from New Zealand and Australia. Fish are deep-skinned to remove the underlying waxy layer, giving very white and firm fillets which look very attractive in retail displays. It has become very popular since its introduction in the early 1980s and supplies have not always been sufficient to keep up with the demand. Some suppliers have repacked smooth oreo dory *(Pseudocyttus maculatus)* as orange roughy. The two fillets look very similar.

Perch

Lake perch *(Perca flavescens)* is a popular fish in inland areas. Larger specimens are filleted. The reddish skin attracts buyers. Ocean perch from the Atlantic *(Sebastes spp.)* is used as a substitute. Lake perch (Figure 5.23, see color insert) is more yellow than the reddish ocean perch from the Atlantic, although the ocean perch loses its reddish tint as it ages: stale ocean perch fillets are unattractively yellow. Pacific ocean perch is covered below under the heading **Snappers and Rockfish**.

Pompano

True pompano *(Trachinotus carolinus)* is a very fine and very expensive fish found mainly in Florida. Permit *(Trachinotus falcatus)* grows much larger and is found in the same waters. The pompano can best be distinguished from the permit by its very long dorsal fin, which reaches back almost to the end of the tail (Figure 5.24). Unfortunately, this long fin may sometimes be broken off (as in the picture), making the fish harder to identify. Permit grow much larger than pompano and are usually sold filleted. Pompano are usually sold whole. Palometa *(Trachinotus goodei)* may also be called pompano (legally) and is distinguished by four dark gray

Figure 5.24. Pompano (top) and permit.

vertical bars on the side. It is slightly thinner than pompano and so yields a little less meat.

There are numerous other species around the world that may be described as pompano or permit.

Salmon

Deliberate species substitutions of one salmon for another are probably rare. Identifying the six different commercial species is almost impossible unless you have the head and fins. Most people tend to look at flesh color and, if it meets their expectations and requirements, assume that they have the right fish. In fact, flesh color is totally unreliable as a guide to salmon species and, as fish farmers learn how to add color to the fish feed, is becoming even less relevant.

For full details of identifying salmon and the differences between the various species, see *Salmon—The Illustrated Handbook for Commercial Users*.

Scallops

There are three major types of scallop sold in the United States. The sea scallop from the North Atlantic is *Placopecten magellanicus*. It is a large scallop, around 20 to 30 count per pound, creamy white. Sea scallops come from domestic and Canadian packers. The bay scallop *(Argopecten irradians)* was found only in the northeastern United States, but is now being farmed in China. The bay scallop is much smaller than the sea scallop, averaging perhaps 70/90 per pound. It is also in chronically short supply and heavily substituted by other scallops from other areas and countries because many people consider that this is the finest of all scallops. The third major domestic scallop is the calico *(Argopecten gibbus)*. This is generally smaller than the bay scallop and is whiter and more elongated. Figure 5.25 (see color insert) shows the three types.

Scallops described as bays are frequently some other species. In a surprising number of cases, calico scallops are the substitute, despite the differences in appearance. Queen scallops from Europe may be thawed and offered as bays. These look very similar. In recent years, as the scallop market has grown, scallops from many parts of the world have been imported. Some of these are good quality material; others, especially from warm waters, are poor.

The best advice on scallops is that you become familiar with the species that you want and be on the lookout for substitutes.

Shrimp

The substitution of one shrimp species for another became headline news when U.S. Customs raided a number of packing plants accused of turning Chinese white shrimp into domestic product. While some packers were thawing, repacking and refreezing shell-on product (see Figure 5.1 as an indication of how similar the different species appear to be), others were peeling the Chinese shrimp and packing the peeled product into containers marked "Product of United States." This practice had been customary and, indeed, was sanctioned by National Marine Fisheries Service inspectors, who maintained that changing the style of the product from shell-on to peeled constituted a sufficient change in the product to affect its provenance. This is likened to importing timber and making furniture from it: the furniture is clearly domestically made, even though from imported raw material. Customs, however, disagreed with this position and decreed that foreign shrimp peeled in the United States should be labeled with its imported origin.

There are at least seventy species of shrimp used commercially and many hundreds more that may appear in trade from time to time. Identifying them requires much skill and, usually, the whole animal—which is seldom possible in a commercial environment.

In most cases, we are reduced to examining the appearance and general characteristics of processed shrimp products to see if they meet previous experience. If they do not, the only resolution may be to have samples tested in a laboratory.

Shell-on shrimp is offered in a range of colors: white, pink, brown, tiger and so on. In general, whites and pinks cost more than browns and tigers, although not all whites cost more than all browns. The Chinese white shrimp caper was unusual in that the origin was comparatively new and the price, because of the novelty of the product, comparatively low. As more buyers become aware of the intrinsic quality of the Chinese shrimp, the price differential is reduced and the temptation to substitute one for the other is lessened.

Generally, look at shrimp cartons to be sure that they are the original ones. Re-strapping, new cartons and changed or deleted markings are all indications that you should look more closely at the nature of the contents.

Peeled and cooked shrimp products are more likely to suffer from substitutions because the opportunities to repack them are greater. A great deal of peeled shrimp is produced in the United States from imported raw material and this is not usually sold as a defined species. Consequently, there is nothing to substitute and you can get on with checking the net contents (see next chapter).

Northern shrimp *(Pandalus borealis)* is a single species which is found around the world in northern waters of the Arctic, Atlantic and Pacific oceans. On occasion, packers may substitute one origin of this shrimp for another, though it is hard to see much harm in this practice. It is seldom possible to distinguish the different origins of a species, even in the best equipped laboratory.

Red Snapper

This is surely the most abused name in the fish business. "Red snapper" has great market appeal. But the name is applied to almost any fish with a reddish skin, including such unlikely and unlike creatures as hybrid tilapia and ocean perch. Under federal rules, the only fish that may be sold as red snapper is *Lutjanus campechanus*. This fish comes from the Gulf of Mexico and is illustrated in Figure 5.26, see color insert. It can be distinguished from other snappers by the rose-red skin on its back, the color becoming lighter towards the underside. It has a longer pectoral fin than other snappers. Nevertheless, it is not easy to distinguish this species.

There are a number of very similar snappers, including *Lutjanus purpureus*, which the FDA allows you to call "Caribbean red snapper," and the Pacific snapper *(Lutjanus peru)*. Many other Gulf and Caribbean snappers have good eating qualities and attractive skin colors. They are excellent fish. However, they are not red snapper and, simply because they are not, they are worth less money.

Identifying the many snappers requires a great deal of experience. Figure 5.27 (see color insert) shows some common snappers. The names are confusing: silk snapper is often called yelloweye. Gray snapper is more often called mangrove snapper in the trade. Vermilion snapper is known as b-liner, sometimes as mingo. Queen snapper is often called ball bat snapper and also misnamed as silk snapper. Figure 5.28, (see color insert) shows the spotted rose snapper, a Pacific species, and the American red snapper. The common and scientific names of species that may legally be called snapper in interstate trade are listed in the Appendix in Table A.1. Note that every one of these names must have a qualifying adjective. The law requires the full name, such as vermilion snapper or mahogany snapper.

Snappers and Rockfish

The name "rockfish" covers a wide group of fish found the length of the Pacific coast. The FDA allows one of these, *Sebastes alutus*, to be called "Pacific ocean perch" (Figure 5.29, see color insert). The others must be called rockfish if they enter interstate trade. The state of California, however, allows 12 species to be called "Pacific red snapper." This name is also permitted in Oregon and Washington, but may not be used for fish of these species which are sold from one of these states to a customer in another.

Rockfish are totally unlike red snapper or other Gulf and Atlantic snappers. Figures 5.29 to 5.31 (see color insert) show some of the more commonly marketed species. The fillets are invariably skinned (real red snapper is always sold with the skin on) and may be labeled as rock cod or ocean perch. Figures 5.32 and 5.33 (see color insert) show three fillets taken from the display case in a supermarket. All three, although labeled differently, are the same fish, Pacific rockfish. Skinless fillets cannot be identified except by laboratory techniques such as iso-electric focussing.

Swordfish

This is a single species, *Xiphias gladius* (Figure 5.34), caught in all warm and temperate seas worldwide. It grows as large as 1,100 pounds. Commercially offered swordfish is usually between 50 and 400 pounds. Mako, green thresher and other sharks are sometimes substituted for swordfish. Swordfish has a rather smoother skin than most sharks. Figures 5.35 and 5.36 (see color insert) show

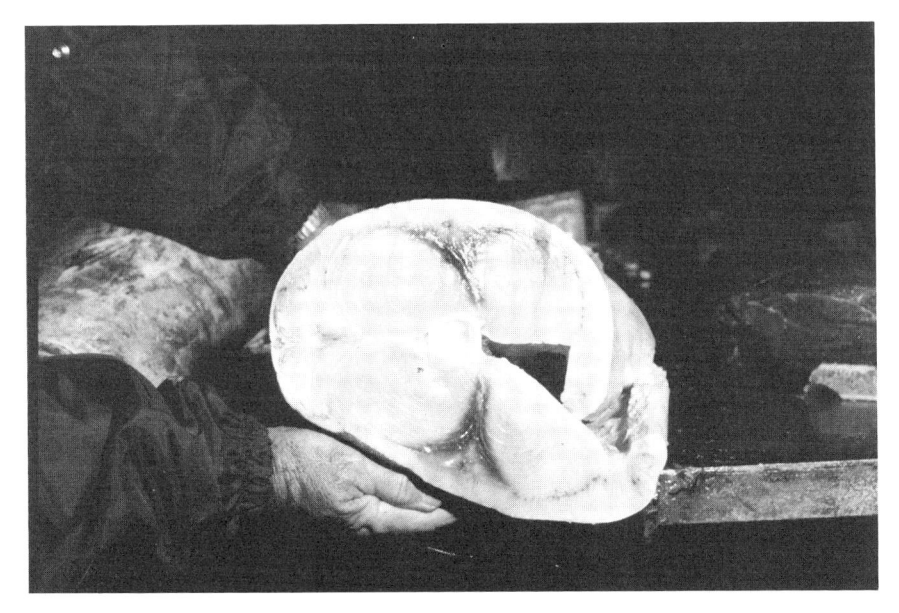

Figure 5.34. Swordfish. Flesh color depends on feed and does not indicate quality.

mako shark with swordfish. Distinguishing between boneless pieces or steaks of swordfish and shark is difficult.

Swordfish flesh varies in color from off-white to pink and brown. There is no quality implication in the color, which is determined by the fish's diet. Some regions have preferences for one color over another.

Chapter 6

Short Weight, Inferior Size and Inferior Quality

Short Weight

Short weight packs are costly for the buyer. They may happen by accident or by design of the packer. They may even occur because of theft of product. Whatever the cause, buyers must protect themselves: you should get the amount of product that you pay for.

Short weight disputes are, unfortunately, quite frequent. But how do you determine the true net weight of a package, especially if it has been frozen? Glaze adds weight, but not product. Many seafoods leach natural moisture: scallops, the most notorious example, if thawed and left to drain will continue to lose natural fluids until they are small shadows of their original size and weight. The objective of any method of testing the net weight of product must be to determine as accurately as possible the weight of product that was originally put into the container.

It is clearly necessary to check weights of product received, but it is also necessary to do this in a careful way which allows for the natural properties of and variations in the product being examined. This is true whether the suspected short weight is accidental—as it often is—or deliberate, as the aggrieved buyer sometimes presumes. This section discusses various ways in which short weight product might be produced and then examines ways to check easily and fairly.

79

Re-marking

The simplest way to supply short weight is to change the weight marked on the box to a greater weight. Because most of the world's countries incomprehensibly use the metric system, a great deal of imported product is packed in units of 2 kilograms. One kilo is 2.2046 pounds. 2 kilos weigh 4.4092 pounds, or 4 pounds and 6.55 ounces (70.55 ounces). Since many 2kg packs, such as blocks of shrimp, are the same size and shape as domestic blocks of five pounds (80 ounces) net weight, a dishonest dealer can change the indicated weight on the 2kg pack to 5 pounds and make an extra profit of 12 percent, because the customer receives 12 percent less product than he is billed for. Figure 6.1 shows a 5lb pack next to a 2kg pack. Sometimes, they are the same size box. In this case, the 2kg box is actually larger than the 5lb container. Note that sometimes 2kg packs are printed for 5lb net weight as well: the box might be packed with 5 pounds of product or 2 kilos, depending on where the packer expects to sell it. Such a box can easily be re-marked.

Shrimp is high priced and particularly vulnerable to this sort of activity. If you ever receive blocks of shrimp (or other seafoods)

Figure 6.1. The 2kg package (top) is actually slightly larger than the 5lb box, because of different glazing methods used.

that have been removed from their inner carton and are simply wrapped in polyethylene, you should suspect strongly that these are 2kg blocks being sold as 5lb blocks. The printed carton has been removed to conceal the stated true net weight.

Because of the weight of glaze that is necessarily included with shrimp and other seafood products, there is no way to check the net weight without defrosting and weighing the contents. The correct way to do this is detailed below (see Chapter Seven).

The problem is not confined to frozen product. Fresh fish is normally buried in ice. Weighing it is messy, time consuming and costly, as the product then has to be re-iced. Nevertheless, it is worth sampling product from each supplier from time to time. Wash off all the ice and then weigh the product. The gross weight of the box, with ice, is no guide whatsoever to the net weight of the contents.

Short Packing

Putting less than the stated amount in a box is another simple way to cheat on the weight of frozen product. Because determining true net weight is not always easy (see Chapter Seven) small shortages are hard to detect. In practice, many good packers deliberately pack a small amount of overweight to allow for any natural shrinkage of seafood when frozen and thawed. The increasing use of highly accurate electronic scales actually makes it easier to cheat because packers can set the target weight at a very small amount under, which is almost impossible to detect. Most packers are honest and have brand names to protect. Short packing is most likely to be found with unbranded and unfamiliar product. It is also more likely from certain origins.

Product that has been repacked into new cartons should always be checked carefully to make sure that the weights are full.

Pilferage

Occasionally, someone will steal product from cartons. Lobster tails, which are individually frozen and bagged and are small, are particularly vulnerable to this. Figure 6.2 shows a 10lb box of lobster tails properly packed. Figure 6.3 shows the same box after

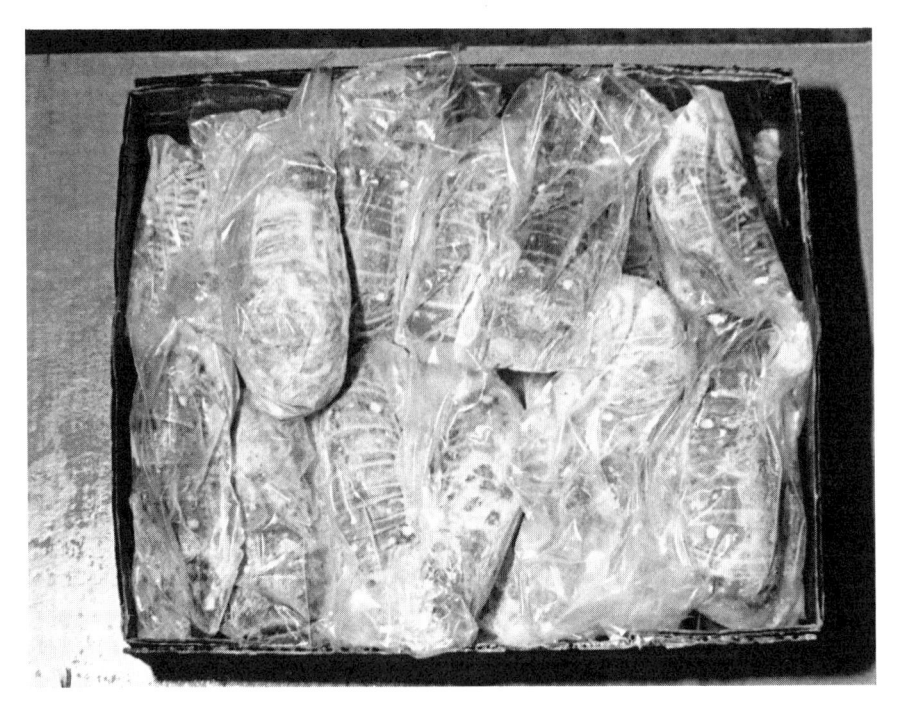

82   Short Weight, Inferior Size and Inferior Quality

Figure 6.2.   Full box of lobster tails.

Figure 6.3.   The same box, minus two tails.

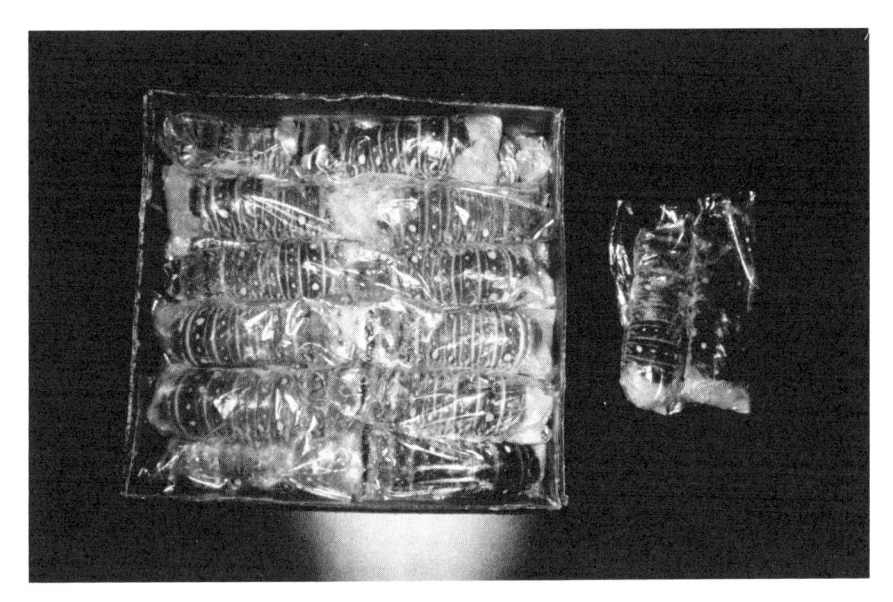

Figure 6.4. Two tails have been removed from this box, without leaving any gaps in the neat arrangement of the tails.

two tails have been removed. As you can see, the box still looks full: it is only by comparing it with the "before" picture that you can see that the pack has been disturbed. Figure 6.4 shows that the same treatment can be applied even if the tails are neatly finger packed. Salmon is less vulnerable because it is larger, but because of the varying weights of boxes of salmon it is sometimes possible to steal one fish from a box without anyone noticing.

Always check that all cartons are sealed and strapped uniformly and with the original strapping (see Figure 1.4 in Chapter One). Damaged or re-strapped cartons are signs of possible pilferage.

Pilferage is not limited to removing a small amount of product from a box: a more determined thief might steal a whole inner box of product by removing it and placing the top layer back in place (see Figure 6.5). This is easy to detect if you are lifting each carton in a shipment, but much harder if you are handling product on pallets or have only one carton of any particular item. Damaged, torn or re-strapped cartons are a symptom: always check the contents of these before you sign the bill of lading from the delivering driver. The greediest thief may steal an entire carton from the center of a pallet, similarly restoring the outer layers of the cartons so that the theft is not immediately noticeable.

Figure 6.5. One complete inner box of rock shrimp was removed and the top layer of boxes replaced.

Glaze

Glaze is another method of supplying less product than stated on the box. Some glaze is absolutely essential, but too much can sometimes translate into a net weight problem. Figure 6.6 shows a lobster tail that has been particularly heavily glazed. However, most excessive glazing is not this blatant and accompanies short packing. Even with lobster tails, cheats are getting more subtle, using injection needles to put water between the shell and the meat (or even into the meat), thus adding quite substantial weight to the individual tails, without any indication of heavy glaze visible on the outside. Some work relating the proportions of lobster tails may, if confirmed, make it easier to spot this type of cheating. Buying lobster tails only from known and trusted suppliers is the best defense.

Excessive Dipping

Phosphates and related chemicals are widely used in food processing. They help to whiten seafoods and to improve the retention of natural moisture, so that scallops, for example, do not lose so much

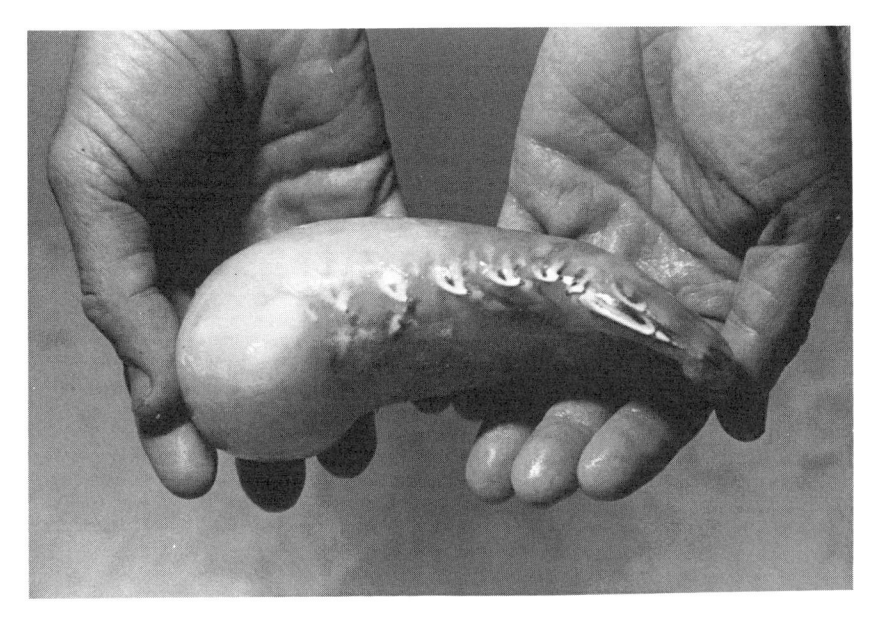

Figure 6.6. A heavily glazed lobster tail.

of their natural fluids, especially after thawing. Abuse of phosphates and other dips can cause some products to absorb additional water, thus increasing the weight. Scallops, shrimp and fish fillets can all be persuaded to increase their weights in this way. Excessive dips can sometimes be noticed because the product has a smooth, slimy feel. If a great deal too much is used, product will not cook properly, but will remain a little translucent. Excessive dips also mask off odors in raw seafoods. The smell is very apparent during and after cooking, though, which is one very good reason why samples should be cooked and eaten.

Inferior Size

Large shrimp cost more than smaller ones. Lobster tails of moderate size cost more than smaller ones and larger ones. Generally, larger sea scallops cost more than smaller ones. Big salmon are worth more than small salmon. This is true for a wide variety of fish and shellfish, although there are, of course, many exceptions.

Shrimp price differentials, even between adjacent counts, may be very large, sometimes in excess of a dollar a pound. In such market circumstances, it is a temptation to the dishonest dealer to sell, say, 31/35 count shrimp relabeled as 26/30 count. Most people cannot readily tell the difference by eye. Figure 6.7 (see color insert) shows six sizes of shrimp, taken directly from the grading machine. Adjacent sizes are very similar and often look the same to the eye. The only way to be sure is to check the counts. Methods for doing this are detailed in Chapter Seven. Similarly, lobster tail prices vary greatly with the size of the tail and smaller tails can often be passed off as premium sizes.

Salmon packers will normally write on each carton the number of fish it contains. If they are 6/9lb fish, you would not expect to get more than 16 in a 100lb carton (the average fish would then be close to the 6lb size). But if one of the 16 fish is, say, 12 pounds, then all or some of the remainder could be under the stated minimum size of the range (6 pounds). Check size consistency as well (see Chapter Seven for methods).

Inferior Quality

Supplying product that is not as good as it is described is another form of cheating and one which is quite difficult to detect. Every product is different and no one person can hope to know all the quality ramifications of each. Poor workmanship is one facet of this problem.

Ratpacking is the practice of putting good quality product on the top of the box, poor quality underneath. Figure 6.8 shows a container of apparently good quality fillets. Figure 6.9 shows that underneath the attractive top layer there are poorly trimmed pieces. It is important to look through an entire container, not simply open it up and accept it on the basis of the top layer. Similarly, the sampling methods outlined in Chapter One should be applied to selecting which containers to sample. If you always look at the top of the first box off the truck, a dishonest supplier will make sure that the first box always contains the best product.

Freshness is as important for frozen fish as it is for fresh, unfrozen product. While rancidity and freezer burn may be obvious even while the fish is still frozen (see above, Figure 3.11), decomposition is not. Decomposition is recognized by smelling the fish. Thaw a piece, or take a core sample and thaw that. If you thaw

Inferior Quality 87

Figure 6.8. A container apparently full of good quality fillets.

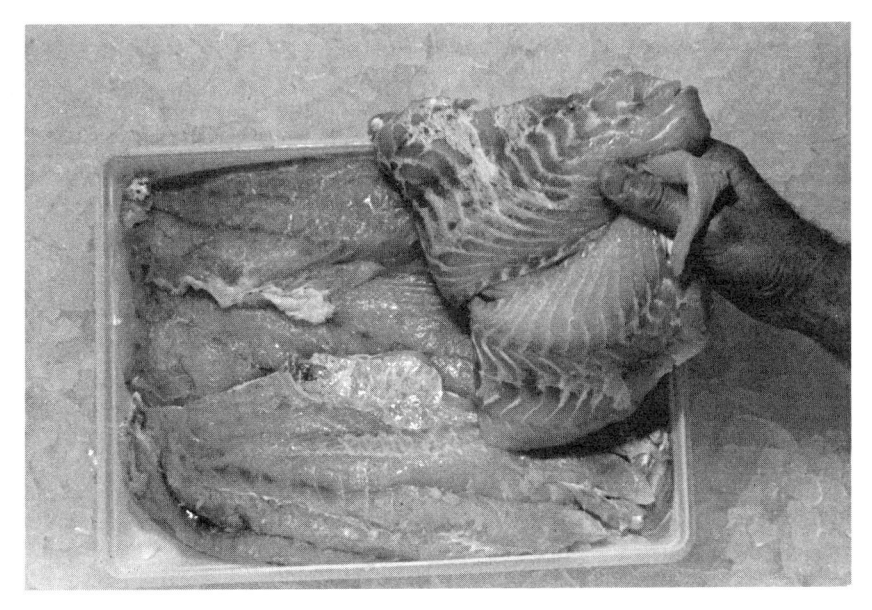

Figure 6.9. The same container showing poorer fillets underneath.

fish under running water, you will wash away much of any off odor it may have. Defrost in a bag or allow it to thaw in a bowl.

Descriptions of quality problems and what to look for with hundreds of seafood products are given in *The New Frozen Seafood Handbook* and in *The New Fresh Seafood Guide*.

Chapter 7

How to Test Product

Standards

There are certain procedures which need to be done in a standardized way to ensure that you get consistent and accurate results. If your tests indicate that a product does not meet agreed specifications, then it is important that your supplier, or an independent inspection service, should be able to replicate your test and your results.

The federal government has developed a number of procedures which apply to seafoods and the best rule is to use these as far as possible and to adapt them when it is not possible. Although at the moment, the administration of fishery inspection is in the midst of intensive political discussion which may well lead to changes, at this time there are two agencies involved: the Food and Drug Administration (FDA), which is responsible under U.S. law for ensuring the safety of all food products; and the National Marine Fisheries Service (NMFS), which operates inspection services for a range of seafood products. NMFS inspection services cover perhaps three-quarters of all the seafoods sold in the United States.

The FDA has defined Standards of Identity for several seafood products. This means that these products can be labeled simply with the description, but must conform to the specifications laid down in Title 21 of the Code of Federal Regulations. Standards of Identity currently exist for oysters (including canned oysters), canned Pacific salmon, canned shrimp, frozen raw breaded shrimp, frozen raw lightly breaded shrimp and canned tuna.

Specifications and ways to test product to check that it meets the specifications are laid down in the Code. It is worth noting that while the oyster standards specify sizes and consistency ratios, the definitions are seldom used by oyster packers, who have mostly developed their own size gradings and descriptions. Nevertheless, the methods used to determine sizes and consistency are valid for any definition.

NMFS has Grade A standard specifications for a wide range of seafood products:

>Whole or dressed fish
>>Frozen headless dressed whiting
>
>Fish steaks
>>Frozen halibut steaks
>>Frozen salmon steaks
>
>Fish fillets
>>General
>>Cod
>>Flounder and sole
>>Haddock
>>Ocean perch and Pacific ocean perch
>
>Frozen fish blocks and products made from them
>>Frozen fish blocks
>>Frozen minced fish blocks
>>Frozen raw fish portions
>>Frozen raw breaded fish sticks
>>Frozen raw breaded fish portions
>>Frozen fried breaded fish sticks
>>Frozen fried breaded fish portions
>
>Crustacean shellfish
>>Shrimp
>>Frozen raw breaded shrimp
>
>Molluscan shellfish
>>Frozen raw scallops
>>Frozen raw breaded and frozen fried scallops
>
>North American freshwater catfish and products made from it

Each standard contains detailed specifications, plus tests to be used to determine whether inspected product actually meets the specifications. Copies of the standards are available from the

National Seafood Inspection Laboratory, P.O. Box 1207, Pascagoula, MS 39568-1207. You can use these standards and tests for the products listed, or adapt appropriate tests for other products that are not listed. For example, you could inspect cod steaks easily by adapting the standard for salmon steaks.

Net Weights of Frozen Seafoods

Because of glaze, discussed in Chapter Three, there are frequent disputes about the net weight of seafoods. To determine whether you have received the correct amount, it is necessary to try to determine how much product was originally put in the box by the packer. This is not always easy, since many products lose natural moisture when they are thawed. To thaw and drain glazed seafoods properly, use an approved method such as that described in the *Code of Federal Regulations*, Title 50, paragraph 265.106. This is based on the only test which will meet legal standards for defrosting seafoods, which is described in *Official Methods of Analysis* from the Association of Official Analytical Chemists. The method described in the Code is as follows:

> Remove package from low temperature storage, open immediately and place contents under gentle spray of cold water. Agitate carefully so product is not broken. Spray until all ice glaze that can be seen or felt is removed. Transfer product to circular No. 8 sieve, 20 cm (8 inches) diameter for product less than or equal to 0.9kg (2lbs) and 30 cm (12 inches) for product greater than 0.9kg (2lbs). Without shifting product, incline sieve at angle of 17–20° to facilitate drainage and drain exactly 2 minutes (stop watch). Immediately transfer product to tared pan (B) and weigh (A). Weight of product = A minus B.

Figure 7.1 shows the simple equipment needed to perform a drained weight test. It is important to defrost frozen product correctly for testing. It is possible to defrost scallops, for example, in such a way that as much as half the weight goes down the drain.

This technique not only meets the need for determining net weights, but can also be used for thawing frozen seafoods to protect the natural flavor and texture as much as possible. Violent thawing, such as under a hard spray of cold water, damages the

92 How to Test Product

Figure 7.1. Equipment for testing drained weight: an electronic scale and a No. 8 sieve.

product and makes it impossible to tell whether it is, in fact, acceptable.

Counts and Uniformity

Seafood is often graded by size or by count rather than by weights. Shrimp, scallops, oysters, clams and many smaller items are usually offered graded by numbers per pound, per gallon or per bushel. For example, a 5lb box of 16/20 count shrimp should contain at least 80 and not more than 100 shrimp. These figures are easy to check: simply defrost a block and count the shrimp. Remember, though that if the thawed product weighs, say 5.5 pounds, there could be a larger number of shrimp while still conforming to the specified size. Similarly, if the block weighs less (for example, 4.4 pounds, there could be as few as 70 but no more than 88 shrimp in the container).

Buyers should accept that complete uniformity is not possible with products such as oysters, scallops, shrimp and fish fillets.

It is really only possible with portion-controlled products made from fish blocks: portions and coated items which are cut from large, standard fish blocks. A reasonable way to check uniformity of scallops is as follows. Start with a sample of, say, a single 5lb box. Remove all the pieces (many specifications will not allow any pieces). Then select the largest 15 percent (12 ounces in the event that there were no pieces) by count and the smallest 15 percent by count from the sample. Divide the total weight of the largest 15 percent by the total weight of the smallest 15 percent. If the answer is less than 2.5, the scallops are very uniform. If the answer is greater than 3.3, the scallops are not uniform. This may be cause for rejection for certain applications where it is important to have evenly sized scallops. This procedure is based on the USDC Grade Standard. For other products, where close uniformity may be unnecessary, a higher number would be acceptable.

NMFS Grade standards for shrimp evaluate uniformity as follows: visually select and weigh the largest 10 percent of the whole, undamaged shrimp in a sample unit. Visually select and weigh not more than 10 percent of the smallest whole shrimp, by count in the sample unit. Divide the weight of the large shrimp by the weight of the small shrimp. The result is the uniformity ratio.

Cooking

Taste testing product is very important. Cooked product reveals taste, texture, odor and, very often, freshness quite dramatically. Since most seafood is eaten cooked, you need to evaluate in the state in which it will be finally judged by your customers.

Although USDC standards for shrimp allow boiling the product with salt, it is generally advisable to cook samples without anything added. The simplest method is to wrap the sample in aluminum foil and steam it on a rack over boiling water for 20 minutes. Alternatively, place the sample in a boilable bag, exclude the air by plunging it in water and sealing the neck, then boil it for 20 minutes.

Appendix

Table A.1. Snapper Names in the United States.	
Scientific Name	*Common Name*
Apsilus dentatus	Snapper, black
Etelis carbunculus	Snapper, ruby
Etelis coruscans	Snapper, yellowstriped
Etelis oculatus	Snapper, queen
Lutjanus analis	Snapper, mutton
Lutjanus aratus	Snapper, mullet
Lutjanus bohar	Snapper, twinspot
Lutjanus buccanella	Snapper, blackfin
Lutjanus campechanus	Snapper, red
Lutjanus colorado	Snapper, colorado
Lutjanus cyanopterus	Snapper, cubera
Lutjanus fulvus	Snapper, blacktail
Lutjanus gibbus	Snapper, humpback
Lutjanus griseus	Snapper, gray
Lutjanus guttatus	Snapper, spotted rose
Lutjanus inermis	Snapper, golden
Lutjanus jocu	Snapper, dog
Lutjanus jordani	Snapper, rufous
Lutjanus kasmira	Snapper, bluestriped
Lutjanus mahogoni	Snapper, mahogony
Lutjanus malabaricus	Snapper, malabar
Lutjanus monostigma	Snapper, onespot
Lutjanus novemfasciatus	Snapper, Pacific dog
Lutjanus peru	Snapper, Pacific
Lutjanus purpureus	Snapper, Caribbean red
Lutjanus rivulatus	Snapper, blubberlip
Lutjanus sanguineus	Snapper, scarlet
Lutjanus sebae	Snapper, emperor
Lutjanus synagris	Snapper, lane
Lutjanus vivanus	Snapper, silk
Macolor macularius	Snapper, midnight
Macolor niger	Snapper, black and white
Ocyurus chrysurus	Snapper, yellowtail
Pristipomoides filamentosus	Snapper, crimson
Pristipomoides macrophthalmus	Snapper, cardinal
Rhomboplites aurorubens	Snapper, vermilion
Symphorichthys spilurus	Snapper, sailfin

SOURCE: FDA *Fish List*

Table A.2. Flounder and Sole Names in the United States.

Scientific Name	Market Name	Common Name
Ancylopsetta dilecta	Flounder	Flounder, three-eye
Arnoglossus scapha	Flounder	Flounder
Aseraggodes kobensis	Sole	Sole, kobe
Aseraggodes macleayanus	Sole	Sole, narrowbanded
Atheresthes evermanni	Flounder	Flounder, kamchatka
Atheresthes stomias	Flounder arrow tooth	Flounder, arrowtooth
Austroglossus microlepis	Sole	Sole
Austroglossus pectoralis	Sole	Sole
Bothus lunatus	Flounder	Flounder, peacock
Bothus mancus	Flounder	Flounder, tropical
Bothus ocellatus	Flounder	Flounder, eyed
Bothus pantherinus	Flounder	Flounder, panther
Clidoderma asperrimum	Sole/flounder	Sole, roughscale
Colistium guntheri	Flounder	Brill, New Zealand
Colistium nudipinnis	Flounder	Brill
Embassichthys bathybius	Sole/flounder	Sole, deepsea
Eopsetta jordani	Sole/flounder	Sole, petrale
Glyptocephalus cynoglossus	Sole/flounder	Flounder, witch/gray sole
Glyptocephalus zachirus	Sole/flounder	Sole, rex
Hippoglossoides elassodon	Sole/flounder	Sole, flathead
Hippoglossoides robustus	Flounder	Flounder, Bering
Hippoglossina stomata	Sole/flounder	Sole, bigmouth
Isopsetta isolepis	Sole/flounder	Sole, butter
Lepidopsetta bilineata	Sole/flounder	Sole, rock
Lepidorhombus boscii	Flounder/whiff	Scaldfish, fourspot
Lepidorhombus whiffiagonis	Flounder/whiff	Megrim
Limanda aspera	Sole/flounder	Sole, yellowfin
Limanda ferruginea	Flounder	Flounder, yellowtail
Limanda limanda	Flounder/dab	Dab, common
Limanda proboscidea	Flounder/dab	Dab, longhead
Liopsetta glacialis	Flounder	Flounder, Arctic
Lyopsetta exilis	Sole	Sole, slender
Microstomus kitt	Sole	Sole, lemon
Microstomus pacificus	Sole dover	Sole, dover
Microchirus variegatus	Sole	Sole, thickback
Paralichthys albigutta	Flounder	Flounder, Gulf
Paralichthys dentatus	Flounder/fluke	Flounder, summer
Paralichthys lethostigma	Flounder/fluke	Flounder, southern
Paralichthys oblongus	Flounder	Flounder, fourspot

Table A.2. (continued)		
Scientific Name	*Market Name*	*Common Name*
Paralichthys olivaceus	Flounder	Flounder, olive
Paralichthys squamilentus	Flounder	Flounder, broad
Paralichthys patagonicus	Flounder	Flounder, Patagonian
Parophrys vetulus	Sole	Sole, English
Pelecanichthys crumenalis	Flounder	Flounder, longjawed
Pelotretis flavilatus	Flounder	Sole, New Zealand lemon
Peltorhampus novaezeelandiae	Flounder	Sole, New Zealand
Platichthys flesus	Flounder/fluke	Flounder, European
Platichthys stellatus	Flounder	Flounder, starry
Pleuronichthys coenosus	Sole/flounder	Sole, C-O
Pleuronichthys decurrens	Sole/flounder	Sole, curlfin
Psettichthys melanostictus	Sole/flounder	Sole, sand
Psettodes erumei	Flounder	Flounder, Indian Ocean
Pseudopleuronectes americanus	Flounder/sole	Flounder, winter/lemon sole
Pseudorhombus arsius	Flounder	Flounder, largetoothed
Pseudorhombus jenynsii	Flounder	Flounder, small-toothed
Pseudorhombus pentophthalmus	Flounder	Flounder, fivespot
Rhombosolea leporina	Flounder	Flounder, yellowbelly
Rhombosolea plebeia	Flounder	Flounder, sand
Rhombosolea retiaria	Flounder	Flounder, black
Rhombosolea tapirina	Flounder	Flounder, greenback
Samariscus triocellatus	Flounder	Flounder
Scophthalmus aquosus	Flounder	Windowpane
Scophthalmus rhombus	Flounder	Brill
Solea vulgaris	Sole	Sole, European
Synaptura orientalis	Sole	Sole, oriental black
Xystreurys liolepis	Sole/flounder	Sole, fantail

SOURCE: FDA *Fish List.*

Resources

This section lists publications that contain helpful information on matters relating to seafood quality and identification.

American Seafood Institute Report. 406A Main Street, Wakefield, RI 02879.

Aquaculture Magazine. P.O. Box 2329, Asheville, NC 28802.

Australian Fisheries. Department of Primary Industry, Canberra, ACT 2600, Australia.

FAO Infofish Marketing Digest. P.O. Box 10899, Kuala Lumpur 01-02, Malaysia.

FDA Consumer. 5600 Fisher's Lane, Rockville, MD 20857.

Food Chemical News. 1101 Pennsylvania Avenue, S.E., Washington, DC 20003.

Marine Fisheries Review. 7600 Sands Point Way N.E., Seattle WA 98115.

Seafood Business. 120 Tillson Avenue, Rockland, ME 04841.

Seafood Leader. 1115 NW 46th Street, Seattle, WA 98107.

Seafood Trend. 8227 Ashworth Ave N., Seattle, WA 98103.

Sea Grant Abstracts. P.O. Box 125, Woods Hole, MA 02543.

Shrimp Notes: A Market News Analysis. Shrimp Notes Inc., 417 Eliza Street, New Orleans, LA 70114.

Bibliography

1978. *Fish Inspection Regulations of British Columbia.* Victoria, British Columbia: Government of British Columbia.

1980. *Official Methods of Analysis of the Association of Official Analytical Chemists*, ed. William Horwitz. Washington, DC: Association of Official Analytical Chemists.

1981. *Guide Book to New Zealand Commercial Fish Species.* Wellington, New Zealand: New Zealand Fishing Industry Board.

1983. *Seafood Sense—A Handbook for Fish Retailers.* New Zealand: Fishing Industry Training Council.

1984. *Manuals of Food Quality Control 5—Food Inspection.* Italy: Food and Agriculture Organization of the United Nations.

1984. *Seafood Retail Training Manual.* Chicago, Illinois: National Fisheries Education and Research Foundation Inc.

1985. *Federal, Food, Drug and Cosmetic Act, As Amended.* Washington, DC: U.S. Government Printing Office.

1985. *Meat and Poultry Inspection—The Scientific Basis of the Nation's Program.* Washington, DC: National Academy Press.

1988. *Fish List, The—FDA Guide to Acceptable Market Names for Food Fish Sold in Interstate Commerce.* Washington, DC: U.S. Government Printing Office.

1988. *Recommended Marketing Names for Fish.* Canberra, Australia: Australian Government Publishing Service.

1988. *Seafood Handlers Manual*. Boston, Massachussetts: New England Fisheries Development Foundation Inc.

1988. *Seafood Safety: seriousness of problems and efforts to protect consumers*. Report to the Chairman, Subcommittee on Commerce, Consumer and Monetary Affairs, Committee on Government Operations, House of Representatives. Washington, DC: United States General Accounting Office.

1989. *Assessing Human Health Risks from Chemically Contaminated Fish and Shellfish: A Guidance Manual*. Washington, DC: United Sates Environmental Protection Agency.

1989 Revision. *National Shellfish Sanitation Program, Manual of Operations Part I, Sanitation of Shellfish Growing Areas*. Washington, DC: U.S. Department of Health and Human Services, Public Health Service, Food and Drug Administration.

1989 Revision. *National Shellfish Sanitation Program, Manual of Operations Part II, Sanitation of the Harvesting, Processing and Distribution of Shellfish*. Washington, DC: U.S. Department of Health and Human Services, Public Health Service, Food and Drug Administration.

1990. *Code of Federal Regulations 21: Food and Drugs, Parts 100 to 169*. Washington, DC: U.S. Government Printing Office.

1990. *Code of Federal Regulations 50: Wildlife and Fisheries*. Washington, DC: U.S. Government Printing Office.

1990. *Interstate Certified Shellfish Shippers List*. Washington, DC: Department of Health and Human Services.

Ade, Robin. 1989. *The Trout and Salmon Handbook*. New York: Facts on File Inc.

Bigelow, Henry B. and William C. Schroeder. 1964. *Fishes of the Gulf of Maine*. Cambridge, Massachusetts: Museum of Comparative Zoology, Harvard University.

Blaufarb, G.P. and E.C. Johnston. 1987. Voluntary U.S. standards for grades of fishery products. In *Seafood Quality Determination*, ed. Donald E. Kramer and John Liston, pp. 665–670. New York: Elsevier Science Publishing Company Inc.

Castro, José 1983. *The Sharks of North American Waters*. College Station, Texas: Texas A & M University Press.

Cheney, Daniel P. and Thomas F. Mumford, Jr. 1986. *Shellfish & Seaweed Harvests of Puget Sound.* Seattle, Washington: Washington Sea Grant Program, University of Washington.

Connell, J. J. 1990. *Control of Fish Quality.* Oxford, England: Fishing News Books.

Crapo, Chuck, Brian Paust and Jerry Babbitt. 1988. *Recoveries and Yields from Pacific Fish and Shellfish.* Fairbanks, Alaska: Alaska Sea Grant College Program, University of Alaska.

Davidson, Alan. 1980. *North Atlantic Seafood.* New York: The Viking Press.

Doré, Ian. 1991. *The New Fresh Seafood Buyer's Guide— A manual for distributors, restaurants and retailers.* New York: Van Nostrand Reinhold/Osprey Books.

Doré, Ian and Claus Frimodt. 1987. *An Illustrated Guide to Shrimp of the World.* New York: Van Nostrand Reinhold/Osprey Books.

Doré, Ian. 1990. *Making the Most of Your Catch—An Angler's Guide.* New York: Van Nostrand Reinhold/Osprey Books.

Doré, Ian. 1989. *The New Frozen Seafood Handbook—A Complete Reference for the Seafood Business.* New York: Van Nostrand Reinhold/Osprey Books.

Doré, Ian. 1990. *Salmon—The Illustrated Handbook for Commercial Users.* New York: Van Nostrand Reinhold/Osprey Books.

Doré, Ian. 1991. *Shellfish—A Guide to Oysters, Mussels, Scallops, Clams and Similar Products for the Commercial User.* New York: Van Nostrand Reinhold/Osprey Books.

Faria, Susan M. 1984. *The Northeast Seafood Book—A Manual of Seafood Products, Marketing, and Utilization.* Boston, Massachusetts: Massachusetts Division of Marine Fisheries.

Gall, Ken. 1986. *Handling Your Catch—A Guide for Saltwater Anglers.* Ithaca, New York: Cornell Cooperative Extension.

Gillespie, Samuel M. and William B. Schwartz. 1977. *Seafood Retailing.* College Station, Texas: Texas A & M University.

Gousset, J. and G. Tixerant. *Les Produits de la Peche: Poissons—Crustaces—Mollusques.* Paris: Ministere de l'Agriculture.

Hart, J.L. 1988. *Pacific Fishes of Canada*. Ottawa, Canada: Department of Fisheries and Oceans.

Hoese, H. Dickson and Richard H. Moore. 1977. *Fishes of the Gulf of Mexico*. College Station, Texas: Texas A & M University Press.

Howell, Thomas L. and LeGrande R. Howell. 1989. *The Controlled Purification Manual*. Boston, Massachusetts: New England Fisheries Development Association, Inc.

Howorth, Peter C. 1978. *The Abalone Book*. Happy Camp, California: Naturegraph Publishers, Inc.

Jensen, Chuck. 1987. *White Fish Processing Manual*. Fairbanks, Alaska: University of Alaska.

Jhaveri, Sudip et al. 1978. *Abstracts of Methods Used to Assess Fish Quality*. Narragansett, Rhode Island: University of Rhode Island.

Joseph, James, Witold Klawe and Pat Murphy. 1988. *Tuna and Billfish*. La Jolla, California: Inter-American Tropical Tuna Commission.

Krane, Willibald. 1986. *Fish: Five-language Dictionary of Fish, Crustaceans and Molluscs*. New York: Van Nostrand Reinhold.

Lamb, Andy and Phil Edgell. 1986. *Coastal Fishes of the Pacific Northwest*. Madiera Park, British Columbia: Harbour Publishing.

Lappin, Peter J. 1986. *Live Holding Systems—A Guide and Reference Manual*. Salem, Massachusetts: Sea Plantations, Inc.

Manooch, Charles S. 1988. *Fisherman's Guide—Fishes of the Southeastern United States*. Raleigh, North Carolina: North Carolina State Museum of Natural History.

Martin, Roy E. and George J. Flick (eds). 1990. *The Seafood Industry*. New York: Van Nostrand Reinhold/Osprey Books.

McClane, A. J. 1977. *The Encyclopedia of Fish Cookery*. New York: Holt, Rinehart and Winston.

McClane, A. J. 1974. *New Standard Fishing Encyclopedia*. New York: Holt, Rinehart and Winston.

Migdalski, Edward C. and George S. Fichter. 1983. *The Fresh and Salt Water Fishes of the World*. New York: Greenwich House.

Mosimann, Anton and Holger Hofmann. 1987. *Shellfish*. New York: William Morrow & Company Inc.

Nelson, Joseph S. 1984. *Fishes of the World*. New York: John Wiley & Sons Inc.

Nettleton, Joyce A. 1985. *Seafood Nutrition: Facts, Issues and Marketing of Nutrition in Fish and Shellfish*. New York: Van Nostrand Reinhold/Osprey Books.

Olson, Robert E. 1987. Marine fish parasites of public health importance. In *Seafood Quality Determination*, ed. Donald E. Kramer and John Liston, pp. 339–355. New York: Elsevier Science Publishing Company Inc.

Organisation for Economic Co-operation and Development. 1990. *Multilingual Dictionary of Fish and Fish Products*. UK: Fishing News Books.

Otwell, W. Steven and John A. Koburger. 1985. *Self-Regulation Guide for Calico Scallop Processing*. Tampa, Florida: Gulf and South Atlantic Fisheries Development Foundation, Inc.

Otwell, W. Steven et al. 1984. *Quality Control in Calico Scallop Production*. Tampa, Florida: Gulf and South Atlantic Fisheries Development Foundation, Inc.

Paquette, Gerald N. 1983. *Fish Quality Improvement—A Manual for Plant Operators*. New York: Van Nostrand Reinhold/Osprey Books.

Price, Robert J. and Pamela Tom (eds.) 1990. *Menu and Advertising Guidelines for California Restaurants, Retailers, and Their Seafood Suppliers*. Davis, California: Food Science and Technology, University of California.

Poissons et Fruits de Mer de France. Paris, France: F.I.O.M.

Rainosek, Alvin P. 1985. *Performance Characteristics of FDA's Sampling Plans for Decomposition of Fish and Seafood*. Pascagoula, Mississippi: National Marine Fisheries Service, National Seafood Inspection Laboratory Training Division.

Robins, C. Richard et al. 1980. *A List of Common and Scientific Names of Fishes from the United States and Canada*. Bethesda, Maryland: American Fisheries Society.

Scott, W.B. and M.G. Scott. 1988. *Atlantic Fishes of Canada*. Toronto, Ontario: University of Toronto Press.

Scott, W.B. and E.J. Crossman. 1985. *Freshwater Fishes of Canada*. Ottawa: Fisheries Research Board of Canada.

Seafood Product Quality Code. Tallahassee, Florida: Southeastern Fisheries Association Inc.

Slabyj, Bohdan M. and Gilles R. Bolduc. 1987. Applicability of commercial testing kits for microbiological quality control of seafoods. In *Seafood Quality Determination*, ed. Donald E. Kramer and John Liston, pp. 255–267. New York: Elsevier Science Publishing Company Inc.

Turgeon, Donna D. et al. 1988. *Common and Scientific Names of Aquatic Invertebrates from the United States and Canada*. Bethesda, Maryland: American Fisheries Society.

Walford, Lionel A. 1974. *Marine Game Fishes of the Pacific Coast from Alaska to the Equator*. Washington, DC: Smithsonian Institution Press.

Wheeler, Alwyne. 1975. *Fishes of the World—An Illustrated Dictionary*. New York: Macmillan Publishing Co., Inc.

Whitehead, P.J.P. et al. (eds.) 1984. *Fishes of the Northeastern Atlantic and the Mediterranean Volume I*. Paris, France: UNESCO.

Whitehead, P.J.P. et al. (eds.) 1986. *Fishes of the Northeastern Atlantic and the Mediterranean Volume II*. Paris, France: UNESCO.

Whitehead, P.J.P. et al. (eds.) 1986. *Fishes of the Northeastern Atlantic and the Mediterranean Volume III*. Paris, France: UNESCO.

Williams, Austin B. et al. 1989. *Decapod Crustaceans*. Bethesda, Maryland: American Fisheries Society.

Williams, Austin B. 1988. *Lobsters of the World—An Illustrated Guide*. New York: Van Nostrand Reinhold/Osprey Books.

Yoshino, Masuo. 1986. *Sushi*. Tokyo, Japan: Gakken Co., Ltd.

Index

Abalone, 57
Ammonia, 18
Appearance, 27
Argopecten gibbus, 74
Argopecten irradians, 74
Arrowtooth, 65
Atheresthes stomias, 65
Atlantic cod, 58
Atlantic pollock, 58, 60

B-liner, 76
Bacteria, 8, 45
Bairdii, 53
Ball bat snapper, 76
Bay scallop, 74
Belly burn, 18
Belly cavity, 18
Belly, 16
Black clams, 58
Blackback flounder, 64
Blue crabs, 50
Blue king crab, 62
Bones, 22, 27
Breaded products, 39
Breaded shrimp, , 40
Brown king crab, 62
Bullets, 27

Calico, 74
California halibut, 66

Callinectes sapidus, 50
Cancer borealis, 63
Cancer irroratus, 63
Cancer magister, 63
Carapace, 49
Caribbean red snapper, 76
Cartons, 3
Cherrystone, 47
Clams, 8, 43, 47, 58
Clumping, 34
Coated, 93
Coating, 41
Cod, 58
Codworms, 21
Cooking, 93
Coryphaena hippurus, 71
Counts, 92
Crabmeat, 50
Crabs, 9, 50, 62
Crassostrea gigas, 46
Crassostrea virginica, 46
Cusk, 60
Cuttlefish, 57

Decomposition, 18, 86
Defrosting, 91
Dehydration, 2, 27, 30, 37
Dipping, 84
Dog whelk, 48
Dolphin-fish, 71
Dressed, 13

Dressed fish, 16, 36
Dungeness crabs, 53, 63

Economic fraud, 55
European Dover sole, 66
Eyes, 15, 17

Fillet, 22
Fillets, 8, 13, 18
Filter feeders, 43
Flakes, 20
Flat lobsters, 70
Flavor, 27
Flounder, 64
Flounders and soles, 64
Fluke, 66
Freezer burn, 2, 37, 86
Fresh, 29
Fresh fish, 13
Freshness, 13, 29, 43, 86
Frost, 33, 39
Frozen, 29

Gadoid, 60
Gaping, 20, 39
Gas, 16
Giant squid, 57
Gills, 16, 17
Glaze, 34, 79, 84, 91
Glazing, 30
Glyptocephalus cynoglossus, 64
Gray snapper, 76
Gray sole, 64
Green thresher, 77
Greenland turbot, 64
Grid, 4
Gut Cavity, 17

Haddock, 58
Hake, 58
Halibut, 66
Haliotis, 57
Hard shell clam, 58
Hippoglossus hippoglossus, 66
Hippoglossus stenolepsis, 66

Homarus americanus, 49
Hoplostethus atlanticus, 72

Ice, 8
Imitation, 41
Imitation crabmeat, 63
Inferior quality, 86
Inferior size, 85
IQF, 2, 30
IQF shrimp, 36
Iso-electric focussing, 60, 77

Jonah crabs, 62

Kamaboko, 63
Kidney, 18
King crab, 62

Lacey Act, 49
Lake perch, 72
Langostino, 67
Lemon sole, 64
Lightly breaded shrimp, 41
Lingcod, 66
Lithodes antarctica, 62
Littleneck, 47, 58
Live molluscs, 43
Lobster tails, 81
Lobsters, 9, 49, 67
Loco, 57
Loins, 27
Lutjanus campechanus, 76
Lutjanus peru, 76
Lutjanus purpureus, 76

Mahi-mahi, 71
Mahogany clams, 58
Mahogany snapper, 76
Mako, 77
Mangrove snapper, 76
Mercenaria mercenaria, 47, 58
Merus meat, 62
Microstomus pacificus, 66
Mingo, 76

Index 111

Mollusc meats, 49
Mucus, 14, 16
Mussels, 43, 47
Mya arenaria, 47

National Shellfish Sanitation Program, 43
Net weights, 91
Net weights of shucked shellfish, 49
Northern lobsters, 49, 67
Northern shrimp, 75
NSSP, 43, 48

Ocean perch, 72, 76, 77
Ocean quahogs, 58
Odor, 17, 27
Ophiodon elongatus, 66
Opilio, 53
Orange Roughy, 72
Oxidation, 38
Oysters, 8, 43, 46

Pacific red snapper, 76
Pacific cod, 58
Pacific ocean perch, 76
Pacific oysters, 46
Pacific pollock, 58, 60
Pacific snapper, 76
Packaging, 30
Palometa, 72
Pandalus borealis, 75
Paralithodes brevipes, 62
Paralithodes camtschatica, 62
Paralithodes platypus, 62
Paralychthys californicus, 66
Parasites, 21, 22, 27
Perca flavescens, 72
Perch, 72
Permit, 72
Phosphates, 84
Pilferage, 2, 81, 83
Pinbones, 13
Placopecten magellanicus, 74
Pollock, 58
Pompano, 72
Portion-controlled, 93

Processing defects, 25
Protothaca staminea, 58
Psetta maxima, 66
Pseudocyttus maculatus, 72
Pseudopleuronectes americanus, 64

Quahog, 47
Quality, 13, 29, 43
Queen scallops, 74
Queen snapper, 76

Rancidity, 38, 86
Ratpacking, 11, 86
Re-marking, 80
Receiving report, 1
Red king crab, 62
Red snapper, 76
Red tide, 43, 48
Refrigeration, 1
Reinhardtius hippoglossoides, 64
Rigor mortis, 20, 39
Rock cod, 77
Rock crabs, 62
Rock lobsters, 67, 69
Rockfish, 76
Round, 13

Salmon, 36, 73
Samples, 9
Sampling, 11, 86
Sampling plans, 9, 11
Scales, 14
Scallops, 48, 74, 79
Scophthalmus maximus, 66
Scrod, 60
Scungili, 48
Sea scallop, 74
Sebastes alutus, 76
Sebastes spp., 72
Seriola lalandei, 71
Sharks, 77
Shellfish, 8, 43
Short packing, 81
Short weight, 79
Shrimp, 35, 56, 74, 80
Shucked molluscs, 48

Index

Sides, 27
Silk snapper, 76
Size consistency, 86
Skeletal structure, 24
Skin, 14, 17
Skinning, 25
Slabs, 27
Slime, 14, 17
Slime sole, 66
Slipper lobsters, 67, 70
Smooth oreo dory, 72
Snapper, 76
Snow, 33
Snow crab, 53
Soft-shell clams, 47
Softshell crabs, 50
Sole, 64
Solea vulgaris, 66
Specifications, 1, 90
Spiny lobsters, 69
Spotted rose snapper, 76
Standards, 89
Standards of Identity, 89
Steaks, 13, 26
Stone crab claws, 63
Strapping, 30
Subcutaneous, 26
Substitutions, 55
Summer flounder, 66
Surimi, 41
Sushi, 21
Swordfish, 77

Tags, 44
Temperature recorders, 2
Texture, 27
Theft, 7
Tilapia, 76
Trachinotus carolinus, 72
Trachinotus falcatus, 72
Trachinotus goodei, 72
Trim, 27
Truck, 1
True cod, 60
Turbot, 65, 66

Uniformity, 92
Uniformity ratio, 93
Vermilion snapper, 76

Wheels, 27
Whelks, 48
Whiting, 58, 60
Whole fish, 14
Workmanship, 86
Worms, 21

Xiphias gladius, 77

Yelloweye, 76
Yellowtail, 71